300 Years of FARM IMPLEMENTS and Machinery 1630-1930

Ronald Stokes Barlow

© 2003 by Ronald Stokes Barlow
Printed in the United States of America
All Rights Reserved

No portion of this publication may be reproduced or transmitted in any form or by any means, electronic or mechanical, including photocopy, recording, or any information storage and retrieval system, without permission in writing from the publisher, except by a reviewer who may quote brief passages in a critical article or review to be printed in a magazine or newspaper, or electronically transmitted on radio or television.

Published by

700 East State Street • Iola, WI 54990-0001
715-445-2214 • 888-457-2873
www.krause.com

Please call or write for our free catalog of publications.
Our toll-free number to place an order or obtain a free catalog is 800-258-0929
or please use our regular business telephone 715-445-2214.

Library of Congress Number: 2002113132
ISBN: 0-87349-632-9

Contents

APPENDIX...202
BIBLIOGRAPHY...206
INDEX..207

America's first farmers..7

Early agricultural publications..9

Clearing the virgin forest...22

Horses, mules and oxen...24

Basic hand tools...26

Evolution of the plow..32

John Deere, blacksmith to manufacturer...35

Harrows, preparing the seed bed...38

Seed planting implements..43

Manure and other fertilizers..49

Cultivating the crop...53

Harvesting machinery..57

Cyrus McCormick and Obid Hussey...59

Hay handling tools...67

Threshing machines replace the flail...70

Silos, storing fodder for winter use...75

Corn pickers, binders, cutters and shellers..76

Eli Whitney, inventor of the cotton gin...77

Dairy equipment, butter churns to cow milkers..79

Potato planters, sprayers and diggers..84

Windmills and their makers..86

Wagons, sleds, carts and carriages..93

Steam engines on the farm...96

Gasoline and kerosene-fueled tractors...121

Stationary farm engines...131

Farm implement catalog reprints, 1892 – 1896...132

LIFE IN NEW ENGLAND, 1770.

INTRODUCTION

The average farmer of the early 1800s was still working the land with simple handmade tools such as those pictured below. With the exception of his heavy plow and harrow, all of these tools could be carried into the field on his back, or tossed into an ox cart for longer transport. However, starting around 1830, a handful of American blacksmiths, inventors and entrepreneurs created a new industry devoted to producing a wide array of implements that would forever relieve farmers from a 3,000 year legacy of back breaking manual labor and increase their production beyond sustenance levels.

In 1834 the primitive hand-held cradle sythe was made obsolete by horse-drawn reaping machines and hay mowers, and by 1850 total annual sales of farm machinery in the United States had reached nearly seven million dollars a year. In 1860 they tripled to twenty-one million dollars, and by the early 1900s sales of agricultural implements exceeded $120,000,000 a year.

A whole new world of steel and cast-iron labor saving gadgets became available after the Civil War. By the 1870s dozens of widely circulated trade catalogs carried hundreds of items for rural dwellers and country storekeepers. Anything from flower seeds to blacksmith's tools could be ordered by mail or railway express. A typical catalog offering ranged from apple peelers to bee-smokers, corn shellers, and bone cutters. Also featured were horse-powered treadmills, feed grinders, cultivators, hay mowers, harrows, seed drills, sulky plows, reapers, and threshing machines.

By 1912 there were 50,000 farm machinery dealerships dotting the landscape from coast to coast. Their colorful showrooms displayed the wares of dozens of manufacturers, and farmers were kept aware of new and improved products through newspapers, magazines, catalogs and county fairs.

The printed advertising of the period conveys a sense of the times that can not be duplicated by any other medium and we have made wide illustrative use of trade catalogs, dealer literature, and farm journals throughout this volume.

Horse-drawn farm machinery will be given its rightful place in history. However, during "The Golden Age of Agriculture", from 1909 to 1914, steam engines provided power equal to seven million horses and mules. The gradual substitution of mechanical horsepower for draft animals on American farms is a fascinating story. In 1830, three man-hours were required to grow and harvest a single bushel of wheat, but by 1900 only ten minutes of labor produced the same results. During most of the 19th century the average farmer struggled to grow enough food to support his family; but by the early 20th century many American farmers were shipping their surplus crops all over the world.

Our 300-year agricultural journey begins with the 1620 landing of the pilgrims and ends with essays on traction steam engines and gasoline farm tractors. In between you will find hundreds of antique engravings of everything from butter churns to farm wagons. We hope you enjoy the ride.

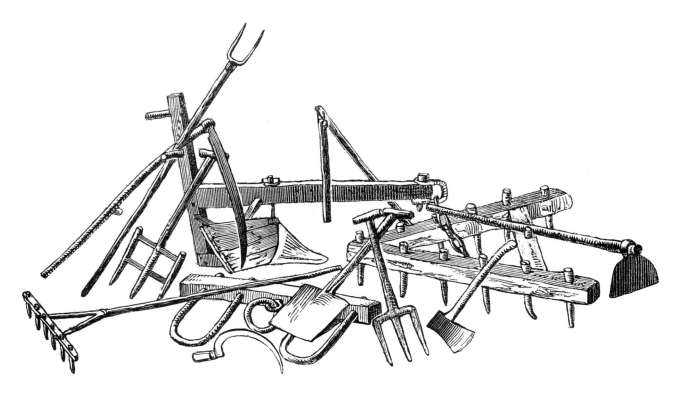

MAKING READY FOR CULTIVATION.

CULTIVATION OF SMALL FRUITS.

THE ABORIGINES.

WHEN our pilgrim forefathers established Plymouth Colony in 1620, they only brought a handful of tools, a few pigs and chickens and no beasts of burden. Cows were finally shipped over four years later, but the common feed source of wild grasses was insufficient and a lot of animals died from malnutrition or exposure.

Most of the accessible land was covered by dense forest and every acre of ground required weeks of hard labor to prepare for planting. No previous experience could have prepared these old-country immigrants for the difficult conditions they were to encounter.

Wolves, bears and hostile Indians killed off pilgrim cattle almost as fast as they multiplied. In 1636, cows sold for as high as thirty pounds sterling, oxen were forty pounds a pair and draft horses were unobtainable at any price. Even the finest plows and harvesting implements would have been useless under these harsh conditions.

Corn, pumpkins, squash, potatoes and tobacco were all foreign plants to early colonists, but the newcomers quickly learned the native methods of cultivation

Indian women were the farm laborers of their race, and even though these ladies owned no metal they fashioned very serviceable mattocks and hoes by tying the dried shoulder blade of a bear, moose, or deer to a short wooden handle.

Indian men planted and cured their own tobacco, it being the only crop they tended with much enthusiasm. The English were introduced to the narcotic weed in 1585 by the Indians of Roanoke, who smoked it in long clay pipes.

Commercial tobacco production in America began in 1607, at Jamestown, where John Rolfe, the husband of Pocohontas, successfully cultivated a large crop of the Spanish variety. By 1619 tobacco was one of Virginia's leading exports and had become a valid substitute for hard cash. Two shiploads of English mail-order brides were actually purchased with pigtails of tobacco. Most of the Virginia settlers of 1620 were bachelors and the treasurer of England sent over 90 girls, each to be exchanged for 150 pounds of tobacco. The new imports were well received by a lonesome male population and a second shipment of brides brought 200 pounds of tobacco per maiden.

The native method of clearing trees from the land was to burn a fire around the base of each tree until its bark was destroyed and the tree died. A squaw with a torch could burn down two or three times as many trees in a month as two white men equipped with iron felling axes. Eventually the new settlers learned to girdle their trees and let them die naturally before burning

THE RESULTING tree ashes enriched the soil, which was prepared for planting corn by digging small holes about four feet apart with a clamshell tool. Into each hole a crab or a fish was deposited along with three or four kernels of corn. Corn was planted in mounds one or two feet high, and beans were later sown at the base of the cornstalks that served as bean poles. Pumpkins and squash were also planted beneath the stalks in order to choke out weeds.

Having few tools, and being averse to manual labor, the noble savage used his brain over brawn at every opportunity. Careful attention was given to the task of protecting crops from destruction by insects, birds and animals. Watch houses, or platforms were constructed in the middle of corn fields, where a squaw or an older child might spend the night, rising early enough to scare off hungry birds, deer and rabbits. Scarecrows were also constructed for the same purpose.

Grain was stockpiled for winter use by Indian women who dug large holes in the earth, which they lined with bark. Corn and beans were sun-dried, or baked on rocks over a fire, and then thrown into these holes and covered with hides and earth. Tightly woven sacks and baskets were also filled and kept in wigwams for daily consumption. More than one starving pilgrim owed his life to the lucky discovery of a hidden Indian granary and the new settlers' hungry hogs delighted in uprooting these underground storehouses

For the next hundred and fifty years there were very few improvements over early Indian farming methods. The deep topsoil, rich with layers of decayed vegetation, did not require careful cultivation. When their land wore out, most settlers would simply move on to another virgin tract and push the natives a little farther west. Eventually the hereditary division of farms and the increased cost of land led to a primitive form of crop rotation. A typical farmer would raise wheat year after year on the same acreage until the soil was drained of nutrients, then he planted corn; when the corn failed he sowed barley, then rye or beans, and so on.

Prior to the Revolution, communication between farmers living more than a few miles apart was very limited. Each landowner planted the same crops, in the same season, as his father and his grandfather before him; and more than likely he wore the same homespun clothes and practiced the same politics and same religion as his forebears. Under these circumstances any innovation on the part of an individual could result in ridicule by the whole community. It is no wonder that new agricultural tools and crop raising methods were slow to be adopted.

The use of fertilizer was little understood in early America. Often so much manure accumulated around a barn that it was necessary to drag it to a new location. Pigs and chickens ran amuck and cows were seldom housed at night, until the extreme cold of winter. Apple orchards were planted but the inferior, and often wormy fruit, was used strictly for cider or its distilled essence, applejack, to which a great portion of the citizenry was sorely addicted.

The Revolutionary War brought men of diverse backgrounds together and newer methods of cultivation were gradually communicated to the nation's farmers. Starting in the 1780's, agricultural societies began to form and scientific papers began to circulate.

SOUTHERN PINE WOODS HOG.

WESTERN BEECH NUT HOG.

IMPROVED SUFFOLK.

BERKSHIRE HOG.

THE CULTIVATOR,

A MONTHLY JOURNAL, DEVOTED TO

AGRICULTURE, HORTICULTURE, FLORICULTURE,

DOMESTIC AND RURAL ECONOMY.

ILLUSTRATED WITH ENGRAVINGS OF

FARM HOUSES AND FARM BUILDINGS, IMPROVED BREEDS OF
CATTLE, HORSES, SHEEP, SWINE AND POULTRY,
FARM IMPLEMENTS, DOMESTIC
UTENSILS, &c.

ALBANY, NEW-YORK:
PUBLISHED BY LUTHER TUCKER, 407 BROADWAY.

OFFICE IN NEW-YORK CITY, AT
M. H. NEWMAN & Co.'s BOOKSTORE, No. 199 BROADWAY,
WHERE SINGLE NUMBERS, OR COMPLETE SETS OF THE BACK VOLUMES CAN ALWAYS BE OBTAINED.

FROM THE STEAM PRESS OF C. VAN BENTHUYSEN.
1850.

By the 1830s there were several agricultural magazines and newspapers in circulation, including *The American Farmer, The New England Farmer, The Genesee Farmer,* and *The Cultivator* (above). On the following pages we have reproduced a dozen examples from an 1850 copy of *The Cultivator*. Among these, McCormick's reaper stands out among the many crude engravings as the most advanced tool of its day.

George Washington and Thomas Jefferson were outspoken advocates for wider dissemination of new farming methods and they both corresponded with leading English agriculturists. The Philadelpha Society for the Promotion of Agriculture was organized by several prominent Pennsylvanians in 1785, and soon other groups were formed by wealthy planters in various states. The Columbian Agricultural Society was organized in Georgetown in 1809, and held its first exhibition in 1810. Prizes were given for the best cattle and sheep, and a spirited plowing contest took place. As time passed, these fairs and exhibits grew in popularity, resulting in substantial advances in farming and stock raising methods.

1850.   PICTORIAL CULTIVATOR.

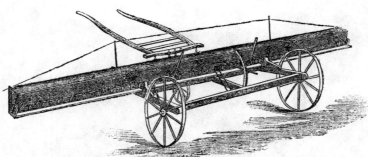

Broadcast Sowing Machine.

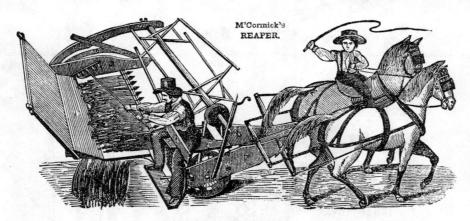

M'Cormick's Reaper.

Pitt's Corn and Cob Cutter.

Sinclair & Co.'s Corn Mill.

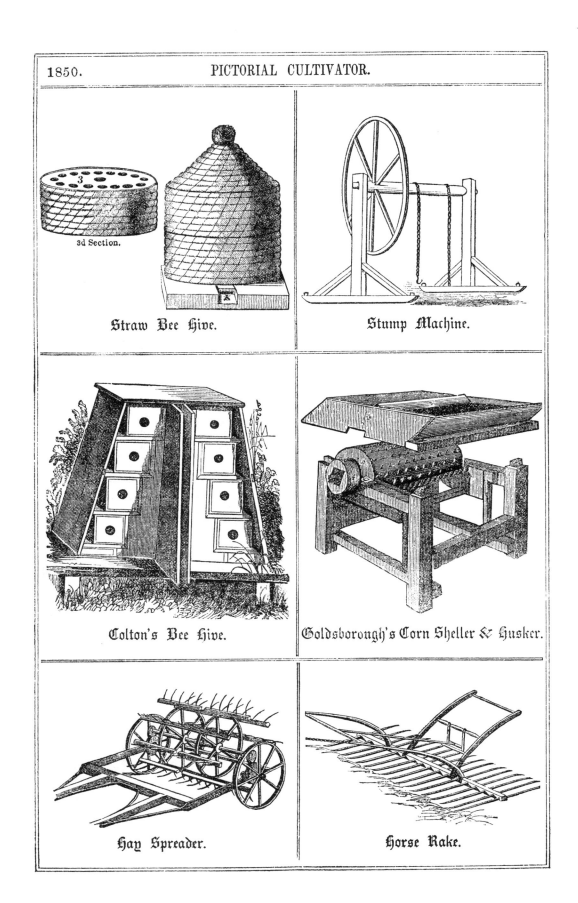

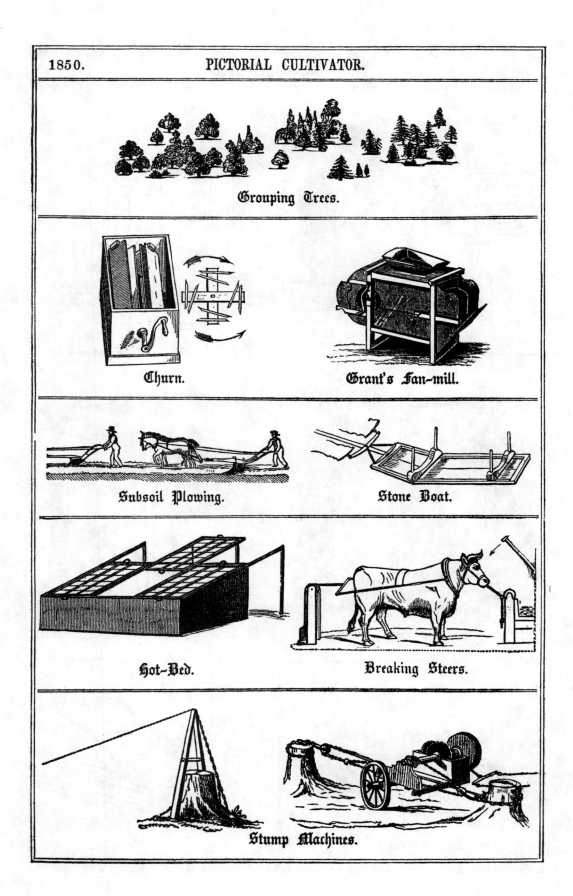

NEW RAILROAD HORSE POWER AND FEED MILL.

THE above cut represents a new Horse Power, recently brought into notice by Messrs. EMERY & Co., of the Albany Agricultural Works. It is on the general plan of the approved endless chain powers sold by them for several years past. The principal difference is in the manner of obtaining and applying the power and motion from the revolving platform to the shaft of the driving pulley.

This Power, as will be readily seen in the cut, has the revolving plank platform, traversing upon its own friction wheels and iron Railroad Track. At the forward end, this platform is supported by its small shafts upon an iron reel, about sixteen inches in diameter—the shaft of this reel extending beyond the sides of the frame work sufficiently to receive a strong converge or internal gear, about twenty-four inches in diameter, as seen in the cut.

The shaft of the driving pulley, (which pulley is three feet in diameter,) is hung in like manner, with the small gear upon one end, operating inside the converge gear before described, and consequently receives an increased motion in the same direction, and carries the driving pulley on the opposite side of the power for driving the Overshot Thresher, without crossing of bands or intermediate gearing. The converge wheel is so arranged as to work on either side of the power, as may be desirable.

This arrangement entirely removes all liability of breakage and wear of links and pinions (heretofore unavoidable,) as the direct stress upon the links working over small pinions is wholly avoided; and they are acknowledged by those using them to run with lighter friction, which it is said enables the power to be operated at a less elevation than by the former mode. The arrangement for tightening the endless platform by means of a joint bolt connecting with the bearings of the reel shaft, is new, and is a very simple and effectual mode of effecting this object, as it may be instantly done by a common wrench without stopping the machine. The platform is considerably longer than usual, avoiding the liability of large or unsteady horses stepping over or off at either end.

The above cut also represents a valuable mill, capable of being driven with this power to good advantage, for grinding food for stock. A considerable number have been sold for several years past, and answer a good purpose. They are cheap, costing but $35, with one extra set of grinding plates,—(new plates costing $2 per set.) and are capable of grinding 600 to 800 bushels per sett, according to the fineness to which it is ground. These are also made and sold by EMERY & Co.

Kendall's Cheese-Press.
Patented July 15, 1843.

The above cut represents an approved cheese press for which the New-York State Ag. Society awarded the first premium in 1848, and is, we learn, generally used in the counties of Herkimer, Oneida, &c., in this state. Its construction is a combination of levers working together, and so arranged as to give any desired amount of pressure. A suspended weight of twenty pounds, being sufficient to give a pressure of ten tons. They can be had of EMERY & Co., of the Albany Agricultural Warehouse. Price $15.

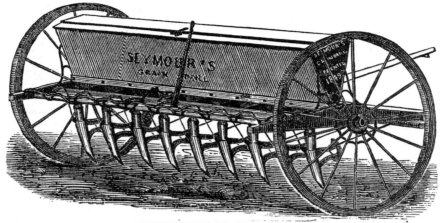

SEYMOUR'S GRAIN DRILL.

This machine is manufactured by P. Seymour, of East Bloomfield, Ontario county, N. Y. It received a premium at the State Fair at Syracuse, as the best grain drill capable of depositing fine manures with the seed. It sows wheat, oats, barley, corn, beans, peas, &c., and is also capable of sowing with the seed plaster, lime and ashes. It can be changed in a few minutes from a drill to a broadcast sower. We have heard this machine recommended by persons who have used it. The price, with nine teeth, is $80; with seven teeth, $70; garden drill $50. For further particulars, see advertisement.

The accompanying cuts represent the appearance of wheat sown both by the drill and broadcast, at the time of ripening. It will be noticed that the height of that sown broadcast is very uneven, while the upright position of many of the heads, indicates that they are light, not being well filled. We have before expressed the opinion that the introduction of the drill system would be an important desideratum.

Wheat sown Broadcast.

Wheat sown with a Drill.

Grain Binders' Wheel Rake.

The above cut represents a labor and time saving implement, used extensively in several states where it has been introduced. It is light, weighing about fifteen pounds. As represented in the engraving, the binder takes the handles and pushes it before him, with the points of the teeth or fingers close upon the ground, and when he has gathered on the fingers a sufficient quantity for binding into a sheaf, he places his foot upon the foot piece, (a.) and by a slight pressure, and by letting go the handles, the fingers and grain are raised above the stubble, when it is readily bound, the binder being required to stoop much less than in the old way of reaching to the ground. When the sheaf is bound and thrown aside, the foot is removed from the foot-piece, (a.) the teeth drop down, and the handles rise ready for the next operation. The wheels are about eighteen inches high, and it is easily pushed before the binder. The width between the wheels is sufficient for the longest grain. It is for sale at Emery & Co.'s Warehouse. Price from 3 to 4 dollars.

Number of Grains in a Bushel of Wheat.—A writer in the *North British Agriculturist* states that the number of grains in a bushel of wheat weighing 62 lbs. is upwards of 630,000.

Sowing Wheat in Drills.—A Scotch farmer estimates the increase of crop from sowing wheat in drills, instead of broad-cast, at an average of one-fourth to one-third.

☞ He who falls in love with himself will find no rivals.

fowls are mostly of the variety called Cochin China, imported by Mr. BAYLIES, of Taunton, Mass. The eggs that were sold were reckoned at fifty cents per dozen, though one dollar was the price charged; but fifty cents per dozen was deducted for the trouble of packing and sending off. The accout stands thus:

Eggs sold—1300,	$27 97
Eggs not sold—581, at 15 cents per dozen,	7 27
Fowls sold,	46 48
Value of fowls on hand over last year at this time,	10 00
	$91 72
32½ bushels of corn and meal, at 75 cents,	24 37
Balance in favor of fowls,	67 35

Wire Fences.

EDITORS CULTIVATOR—I regret to notice in the January number of the Cultivator, an article against *Wire Fences;* and to show that the writer is "reckoning without his host," I will state that No. 10 wire, which is the finest used, can be bought for 5½ cents per pound; and that Mr. ELLET, the constructor of Wire Suspension Bridges, in a report, states that a single strand of this No. will sustain 1500 pounds. A fence of wire of this No. may be made for 50 cents per rod; and in case of the "plunging of heavy cattle against it," they would probably meet with the resistance of three of the wires, and an animal in breaking them must employ a force equal to about 4500 pounds; and the writer alluded to thinks that when made even of wire so large as to cost $2 per rod, it will be "frequently broken" in that way. Now according to the above calculation, they must be mighty cattle, possessing "a power" of strength and force fully equal to the iron-horse called Loco-motive.

Any one who has observed the effect of wind in swaying and breaking off the wires of a well-constructed wire fence, must admit that the objection is at least a windy one.

Now although I have had but about 40 rods of experience in making this kind of fence, and which cost me less than 50 cents per rod, I am fully of the opinion that it is worthy of the attention of every farmer who has not "stones on his fields which he wishes to get rid of;" for I know of no locality where a farmer can erect a rail or board fence for much less than $1 per rod; but I can refer you to many large sections where every farmer will tell you his rail and board fences cost him nearly double that sum, and is yearly increasing as our forests and wood lots are decreasing, and are enhanced in value. That a cheaper substitute for field fences is loudly called for, no man can dispute. It is true that the American farmer is at greater expense for the support of his fences than any other farmer in the world. It is by far the heaviest drawback upon the profits of the farmer, that he is obliged to contend with. Indeed it stands like a lion in the pathway of many great improvements which long ago would have been made, and which every farmer yearly sees and is desirous of making, but is reluctantly compelled to turn from, and apply all the means in his power to his decaying fences of wood. And how, I ask, are our western prairies to be fenced, where in many instances, not a tree is to be seen or stone found for leagues; and it has been found that embankments of earth there will not answer. In consequence of the great cheapness of iron, many have been led to experiment with it in the form of wire as a substitute for wood fences, and I see nothing against its success, after, perhaps, some better method shall have been contrived for straining, fastening and uniting the strands; and should it come into general use, it would add immensely to the great iron manufacturing interest of the country. Three and a half feet high is sufficient for sheep, and 5 feet for cattle and horses; and I should be pleased to give such information as I am possessed of with regard to constructing it.

And now I earnestly suggest to every reader of the extensively circulated Cultivator who has had any experience or observation respecting Wire Fences, to transmit their views to this journal, together with a minute method of constructing it, and I doubt not that an opinion may at once be formed with regard to the propriety of its general adoption.   A. B.

The original signature of the above communication having been that of a well-known and regular correspondent of *The Cultivator,* we have thought it proper to substitute another in its stead. We shall be glad to receive from A. B., and from any other of our readers who have erected wire fences, their views as to the best method of constructing it, its cost, and probable permanence.—EDS.

Products of Labor and Capital.

The Report of the Commissioner of Patents for the year 1848, makes the following estimate of the products of labor and capital in the United States for that year:

	Quantities.		Value.
Wheat,	126,364,600	bushels,	$145,319,290
Indian Corn,	583,150,000	"	344,058,500
Barley,	6,222,050	"	4,044,332
Rye,	32,951,500	"	21,418,475
Oats,	185,500,000	"	64,925,000
Buckwheat,	12,523,000	"	6,266,500
Potatoes,	115,475,000	"	32,342,500
Beans,	10,000,000	"	10,000,000
Peas,	20,000,000	"	17,500,000
Flaxseed,	1,600,000	"	1,920,000

The prices of these articles per bushel, are thus estimated:

Wheat,	115 cents.	Buckwheat,	50 cents.
Indian Corn,	59 "	Potatoes,	30 "
Barley,	65 "	Beans,	100 "
Rye,	65 "	Peas,	87½ "
Oats,	35 "	Flaxseed,	120 "

It will be seen that Indian Corn is estimated at the immense sum of more than $344,000,000—while the gross amount of the Wheat produced was little more than $145,000,000.

	Pounds.	Value.
Tobacco,	218,909,000	$8,756,360
Cotton,	1,066,000,000	74,620,000
Rice,	119,199,500	3,575,985
Sugar,	275,000,000	13,750,000
Silk,	400,000	800,000
Hops,	1,566,301	140,667
Beeswax,	789,525	165,800
Honey,	23,665,750	2,368,575

The prices are thus given:

Tobacco,	4 cents.	Silk,	$2 00
Cotton,	7 "	Hops,	9
Rice,	3 "	Beeswax,	21
Sugar,	5 "	Honey,	10

The quantity of molasses is estimated at 9,600,000, which, at 28½ cents, realized $2,736,000.

Wine, 500,000 gallons, which, at $1, brought $500,000.

The annual value of pasturage is put down at $60,768,136.

The value of the residuum of crops, such as straw, chaff, &c., $100,000,000. Of manure, $60,000,000. Product of orchards, $9,071,130. Of gardens, $45,000,000. Nurseries, $741,917.

Butchers' meat, including mutton, beef and pork, $146,597,360.

Hides, felt and tallow, $20,000,000.

Neat Cattle, $4,401,470. Horses, mules and asses, $8,129,350. Poultry, $11,680,512. Eggs,

$5,431,500. Live geese feathers, $1,000,000.

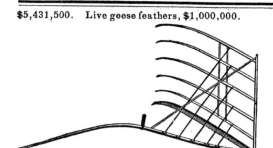

Our Cradles have taken the first premiums at two New York State Fairs, and are considered the best in use.

The great encouragement we have received from dealers and agriculturists, has induced us to greatly enlarge our business, and we hope by strict attention, to merit a further patronage.

Orders will be thankfully received, and receive prompt attention. I. T. GRANT & CO.
Junction P. O., Rens. Co., 8 miles north of Troy

Things in Virginia.

EDS. CULTIVATOR—There can be no question but a large number of our farmers would not hesitate to spare a dollar or more, to see some trifling amusement—a "bear dance," or a cock-fight; and yet are indifferent towards an agricultural paper, which like a true friend, sticks by them, ever ready to point out the poor and wet parts of the farm, saying—"drain, manure, plow deep, harrow fine, seed well, read and observe."

Allow me to make a suggestion. Suppose every member of an agricultural club or society, were to make a report of all his operations at the end of every year, stating the number of his family, the number of laborers and the number of those unable to labor, the number of stock of all kinds, the food consumed by all, the work for each lot or field and the number of acres in each, the kind of crop, the yield, the amount of money received and spent, the state of the weather for each month, with such remarks as suggest themselves—would it not be useful?

Our crops last year were fair, except fruits, of which we had none. I did not see a good peach or apple the whole season. The whole fall and December were very pleasant, till the night of the 30th ult., when it commenced raining, then hailing, and ended in snow, which fell six inches deep. The thermometer was down to 12° on the first of January. J. BUNCH. *Chuckatuck, Nansemond Co., Va., Jan. 1, 1850.*

Mode of Planting Corn.

EDS. CULTIVATOR—We have in this section a method of planting corn which may be new to some of your readers.

After the ground is furrowed one way, one man commences furrowing in the other direction. A boy or man follows and drops the corn. Then another, provided with an implement something like the common shovel plow, with a square piece of iron about the size of a common hoe screwed fast to the end of it, follows the dropper and covers the corn, by letting the iron scrape up the dirt from the bottom of the furrow, and deposit it immediately on the hill. As soon as the corn is covered up, the planter is dropped again for another hill, &c. This method here, in our new fields, where stumps are very thick, is a great saving of time. Two men, and a boy 12 years old, with two horses, can furrow out one way, and plant eight acres in a day. This I know to be true, for I helped to plant one of my fields of eight acres in this manner last spring. The shovel-plow, with a square piece of iron on the end of it, will answer the purpose. W. R. WEBB. *Vienna X Roads, Clarke Co., Ohio, Dec. 27, 1849.*

Profits of the Dairy.

EDS. CULTIVATOR—We have many times noticed in your paper, statements made by different writers of the profits of a dairy. Below we give you an account of the proceeds of our dairy for the year 1849, from forty-one cows, six of which were heifers, having their first calves the same season:

INCOME.
41 calves, at four weeks old, $4 each	$164 00
3,748 lbs. cheese, at 9 cents per lb.	337 32
6,569 lbs. butter, best quality for table use, at 20 cts per lb.	1,313 80
6,570 gallons, or 18 galls. per day, new milk, used on table, never skimmed, at 3 cts. per quart	788 40
For manure	200 00
Total amount	$2,803 52

EXPENSES.
10 tons wheat bran, or ship stuff, at $10 per ton	$100 00
600 bushels beets at 1s per bushel	75 00
62 tons hay, at $8 per ton	496 00
26 weeks pasturing for 41 cows, at 1s per week each	333 25
Slops from kitchen during the year	15 00
Nett expenses	$1,019 25
Total amount	$2,803 52
Deduct expenses	1,019 25
Leaving a balance of	$1,784 27

Making an average to each cow, of butter............160 lbs. 3½ oz.
" " cheese............ 91 " 6½ oz.
" " milk............160 galls.

The milk, it will be understood, is that which is used on table by boarders, never skimmed.

Add manure and calves, and the total amount for each cow is $68 37
Deduct expenses... 24 86

Nett profit to each cow.............................$43 51

Made of butter in the month of October, 1849, 1st week ....180 lbs.
" " " 2d " ....201 "
" " " 3d " ....191 "
" " " 4th " ....187 "
" " " 5th ½ ....173 "

Total in October.................................932 "

We prefer putting our cows in the stable while milking, at all seasons of the year. This affords an opportunity of *messing* twice a day, and is done regularly at time of milking, believing it the best time. Wheat bran, or shorts, mixed with slops from the kitchen, or dairy, make a good feed for milch cows.

Some think it quite objectionable and very unnatural for cows to eat or drink whey and milk, but we see no good reason for such objections.

We have practiced for some years feeding our cows the whey and skimmed milk from our dairy, mixed with wheat, buckwheat, or rye bran, and have never seen any injurious effects whatever—but, on the contrary, believe it to be very beneficial, and productive of good sweet milk and butter.

It is very necessary for milch cows to be well supplied with good pure water, especially in the winter season when fed on dry fodder. We make a practice of watering our cows twice a day, morning and night. This is given them in the stable, where they can drink at leisure, sheltered from cold and storm. *New Lebanon, Shaker Village.* Family of JONATHAN WOOD and EDWARD FOWLER, numbering 130 persons.

A change of fortune hurts a wise man no more than a change of the moon.

A false friend and a shadow attend only while the sun shines.

Fools make feasts and wise men eat them.

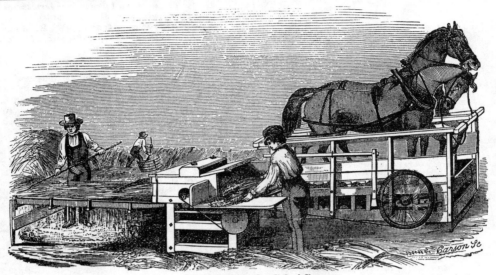

EMERY & CO.'S
LATEST IMPROVED RAILROAD HORSE POWER, AND OVERSHOT THRESHING MACHINE AND SEPARATOR.

THE above cut represents this most useful machine, with the LATEST IMPROVEMENTS, ☞ *For which Patent is secured,* embracing some of great value and importance—which have suggested themselves from time to time as the various kinds made and sold by us have become worn, used and failed.

The most important of these consists principally in the mode of applying the power and motion from the endless platform to the shaft of the main Driving Pulley, and obtaining the necessary motion for the OVERSHOT THRESHING MACHINE, without crossing bands or intermediate gearings, and at the same time dispensing with the small pinions and cogs on the links of the endless platform,—thereby combining GREATER STRENGTH and DURABILITY with LIGHTER FRICTION, without the liability of breakage of links, or the wearing of links and pinions,—(no small item in the expense of repairs in most other kinds of powers in use.) The farmer or mechanic is enabled to perform a greater amount of work, or to operate with less power or elevation, as best suits his wishes.

Having been long engaged in the Manufacture, Introduction, Sale, &c., of the various kinds of Horse Powers, for different purposes, and at all times adopted such improvements as from observation and experiment have seemed necessary and desirable, we feel confident that in this Power, as now manufactured, all that can be desirable, is found to a greater extent than any heretofore sold by us, or with which we are acquainted. They were introduced to some considerable extent last season, and wherever used side by side with the most approved Powers of other kinds, have given unqualified satisfaction, and been preferred.

trouble, always be kept tight. The speed of the Power is such that a larger pulley is used on the Thresher than on most others—driving stronger, with less liability of slipping of Bands, which last are made of Vulcanized India Rubber. The Separator makes a complete separation of Grain from the Straw, leaving it in the best condition for the Fan Mill; thus saving the labor of several men, and doing the work better.

Fan Mills of various sizes, for Hand, or fitted to be driven by the Power, at same time of threshing. Also, Saw Mills in complete order.

The Double Horse Power is capable, with 3 or 4 men, of threshing from 125 to 200 bushels of Wheat or Rye, and the Single one from 75 to 100 bushels, or double that quantity of Oats per day. They are warranted to perform as above, or may be returned to us or our Agents, of whom they were purchased within 3 months, and the purchase money refunded.

They may be had in Rochester, Buffalo, or any of the principal ports on the lower or upper lakes, by adding transportation.

Good agents will attend to the sale of them in those places.

The prices will be, for Single Powers,..........$85 00
" Thresher and Separator,............ 35 00
" Bands, Wrench, Oil Can, extra pieces, 5 00— $125 00
Best Double Machines, Complete, ($25 more on,) 150 00
Fan Mills, from...................... $22 to $28
Saw Mill, complete................... $35
Also "Wheeler's" Machines, improved this season,
Single Setts, complete,.........................$120 00

EMERY'S NEW THRESHER AND CLEANER.

The annexed cut represents a thresher with an apparatus attached to it for cleaning or winowing the grain. It was got up by Messrs. EMERY AND CO., of this city. They have tested its operation throroughly during and since the late harvest, and we learn that it gives entire satisfaction. It cleans the grain ready for market, without waste, as fast as it is threshed. The cost of the cleaning apparatus is about $30 making the cost of the thresher and cleaner, $75.

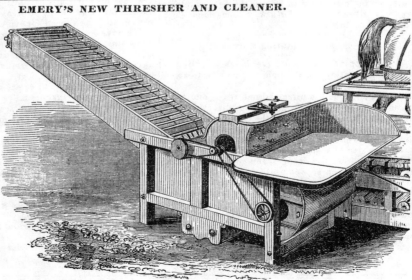

NEW-YORK STATE PREMIUM PLOWS.

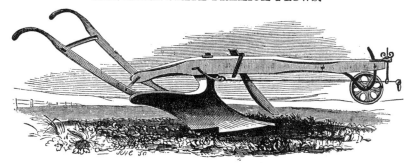

Prouty & Mears' Centre Draft, No. 25:
Which received the first premium for sandy soil.

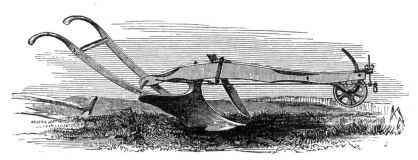

Prouty & Mears' Centre Draft, No. 30:
Which received the first premium for stiff soils

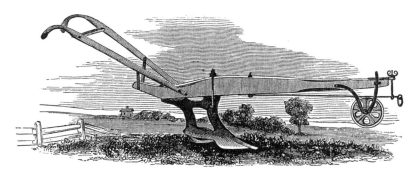

Prouty & Mears' Subsoil Plow, C:
Which received the first premium.

REPORT OF TRIAL OF PLOWS.

In our last we gave cuts of four of the plows which received premiums at the trial by the New-York State Agricultural Society, in June last. Herewith we give cuts of the other premium plows, except the one which received the second premium for sandy soil, which has not yet been received. The report appears to be received with general approbation. JOHN DELAFIELD, Esq., in a late letter, observes: "The plow report is admirable, and seems to me a true and sure standard for future trials. Our agricultural artisans have long exhibited a knowledge of the principles of force and motion, but in the plow they have not reached, perfectly, their application for our most economical use. The present report will bring them in close contact with our farm wants, and their skill will, assuredly, ere long, give us all we desire."

CLEARING LAND FOR CROPS. Various methods of removing native growth were employed in different regions of the country. Axes, fire, horses, oxen, stump-pullers, gunpowder and eventually, dynamite, were all utilized in preparing the ground for plowing. A witness to the clearing of trees and brush from four hundred acres of New York land in 1804 wrote this account: "The job was contracted out to nine men and took almost a year to complete. Great elms, maples and mighty oaks were felled and piled up for burning during the dry months. Smoke from the fire could be seen for fifty miles and the intense heat drove men and cattle into the river for relief and also spoiled fields of green oats a great distance away. However, the newly cleared land produced a crop of ten thousand bushels of wheat that first season."

In Pennsylvania and other areas settled by German immigrants the largest trees were killed by girdling their bark and waiting for them to rot, rather than spending too much time and energy at the end of an axe handle. In New England and in the Western states, the underbrush was cut and piled in heaps for burning, then the larger trees were felled and converted to logs. The smaller trees, and the limbs from the larger ones, were cut into twelve-foot lengths and allowed to dry for several months before burning. Oxen then hauled the logs to one or two locations where they were piled up and burned, along with any of the smaller stuff that had not been consumed by the earlier inferno.

The resulting ashes were put into large tubs with perforated bottoms, called *leeches*, to which a small quantity of salt and lime were added. Boiling water was poured through the ashes and captured in a container below. The resulting black liquid, called *ley,* (lye) was cooked in iron pots over a fire until it evaporated and left a residue called *black salt*, which was then converted to *potash* by applying intense heat to the iron kettles. Very few settlers had the time or equipment to complete the entire potash making process. Most preferred to sell their evaporated black salt to local storekeepers, who then completed the manufacturing process before packing the potash in tight barrels for shipment to large cities where it was inspected and graded before payment was made to the merchants.

Potash (potassium carbonate) was used in soap making, wool processing and glass making. In England, where more than a million people were employed in the woolen industry, this new source of potash from the American Colonies was more than welcome, as previous supplies of the chemical had been imported from Europe at much higher prices.

Even common wood ashes proved to be a lucrative cash crop for the impoverished pilgrims, who shipped a boatload from Jamestown back to England in 1608. By 1670 the area which is now Maine and New Hampshire was making more money on the export of animal fats and wood ashes than on all other crops combined. On the western frontier a smart settler could often pay for the initial clearing and plowing of his land with the ashes from his first "Burning over".

Stumps uprooted by ox teams made formidable fences when laid side by side with their wide root systems facing inward like giant cartwheels. Barbed wire was not invented until 1873, but wood was abundant and a serviceable split rail fence could be built by two or three men at the rate of a mile a week. Stone walls, made from land clearing, and hedgerows of buckthorn, or osage orange, rounded out the early settler's inventory of land partitioning materials.

It is interesting to note that fence rails were cut to a standard eleven feet, which was exactly one-sixth the length of a standard surveyor's chain. This made it easy calculate boundaries without calling in a professional. Two rail-lengths were also the standard width of most early roadways.

ANIMAL POWER on the farm was provided by horses, mules and oxen. One workhorse was needed for every 25 to 30 acres of cultivated land. In a typical nine or ten hour day a sturdy draft horse might travel sixteen or eighteen miles while plowing two acres of farm land.

Man's use of horses dates to before written history. Horses were ridden by the ancient Babylonians as early as 2000 B.C., and from there were transplanted to Egypt about four hundred years later. Excavations in China indicate that horses were domesticated in the Orient about 4,000 years ago.

Until about 350 years ago horses were used almost exclusively for riding; their widespread replacement of oxen as draft animals in Europe and North America is a fairly recent development, dating from the mid 1800's. However, the Romans had developed padded horse collars and harness gear for pulling plows as early as the 12th century.

The so-called wild horses found on the plains of Texas and the western prairies were descendants of Spanish stock abandoned by Hernando De Soto after the failure of his expedition in 1543. Coronado also included horses in his 1540 journey from Mexico to what is now the boundary between Kansas and Nebraska. All of these large Spanish horses were of Arabian descent.

Six mares and a stallion were imported to Jamestown, Virginia, in 1609, and in 1630 several horses were introduced into the Colony of Massachusetts Bay, from Leicestershire, England. The Dutch West India Company imported horses from Flanders into New York in 1682, and it is thought that the Conestogas – a very large and heavy breed of wagon pullers – descended from these animals and English stock. The French, who settled Illinois in 1682, brought along their Canadian ponies which were allowed to range freely in the region. From these sources came the "native" horses of the United States. These animals were later modified by the infusion of imported thoroughbred blood.

There were five chief purposes for which horses were raised: (1) for speed, as trotters and runners; (2) for sport or fashion; (3) for family driving; (4) for farm work; (5) for draft purposes, usually in cities. The first three classes were of the same general type, being smaller and more active than an 1,800-pound draft horse.

Surprisingly no one attempted to develop a breed of horses specifically for farm use. Big city draft horses, like Belgians, Percherons, and Clydesdales, were generally considered too large for agricultural work. However, after 1918, when trucks began to replace them on city streets, they were successfully employed to draw heavy plows and other large pieces of farm machinery.

AMERICAN AGRICULTURIST

CAPTURING WILD HORSES. — *Engraved for the American Agriculturist.*

It is generally admitted that the wild horses of our far Western plains descended from European stock. The term *Mustang*, applied to the wild horse in both South and North America, is a corruption of the Spanish *mesteño* (from *mesta*, pasture). In Mexico, and the portions of our territory bordering upon that country, the Mustang is still largely employed, though it is being rapidly replaced by better breeds. Large herds of Mustangs are found on the broad plains of the South-west, and their capture is still followed by both Mexicans and Americans. The usual method of capture is by the lasso, though sometimes, in favorable localities, they are taken by driving into an enclosure. The skill with which the Mexican throws the lasso, is often a subject of comment, but many Americans are quite as expert in this very useful accomplishment of the herdsman. In lassoing a wild horse, or other animal, the horse ridden by the hunter plays a most important part. The saddle is of great strength, and secured by girths correspondingly strong. One end of the lasso (usually 30 or 40 feet long), is made fast to the horn of the saddle, and when the rider makes his cast, which is done at a full gallop, and the noose falls upon the neck of the wild animal, the horse seems to take as much interest in the capture as the rider himself. When it is necessary to check the captive, the well-trained horse braces himself for the tussle, and renders essential aid. The writer traveled for some months in the region of wild horses, and though "sign" of various kinds was frequent, and our interest was greatly hightened by the wonderful stories told of them, not a drove, not even a single *caballo*, could we see, until at last the party concluded the wild horse to be a myth, and it was the subject of many a joke. Early one morning, a few days out from Corpus Christi, Texas, as we reached a crest of one of the great swells of the prairie, there were wild horses at last, and enough to make up for any former deficiency! As far as the eye could reach, were horses in every direction, in groups and droves—hundreds—no doubt thousands were in view at one time—a sight never to be forgotten! When we first saw them they were quietly grazing, but they soon saw us, and we then found that the wild horse was far from a "myth," and the meeting with them was very unlike a "joke." Perhaps the white covers of our train of a dozen or more 4 and 6-mule wagons attracted their curiosity, at all events they seemed to wish a nearer inspection, and down they came, in squadrons, regiments, and divisions, circling at full speed, nearer and nearer, their ample manes and tails giving an air of surpassing grace. We did not have long to admire the wonderful beauty of this scene. The mules, from their well-known sympathy with horses, became fairly excited, and soon were unmanagable. It was impossible to keep them in line, and though by firing at the wild horses with rifles and pistols, they took alarm and began to make off, the trouble then was that the mules would make off too! Then followed a "circus," compared to which the "Greatest Show on Earth" is a tame affair. Over 50 frantic mules, and a dozen or more equally frantic Texan teamsters, the mules with braying, trying to go in one direction, and the teamsters, with quite the reverse of praying, urging them to go in another, all emphasized by the *staccato* of the drivers' whips, might not be called a "Great Moral Entertainment," but was a very lively one. In the midst of it all, one

HORSE BREEDERS of the 1870's did quite well. An agricultural writer of the period noted: "Our wagons are now of lighter construction, our plows run easier, our lands are freer from rocks and stumps, and quick, hardy horses often take the place of oxen and of the heavier and much slower horses of fifty years ago. The purposes for which horses are now wanted are very different than what they used to be. Most people want a trotting horse to go off easily at eight, ten, or twelve miles an hour. A demand very soon creates a supply, and the farmer who breeds horses will supply the tastes of the community. Few farmers will keep a horse for general farm work that will bring from two to five hundred dollars."

Farmers in the middle colonies began to shift from oxen to horses for plowing and wagon pulling in the early 1700's. By selective breeding these farmers developed stallions from 13 to 15 hands tall. Pennsylvania Dutch farmers concentrated on producing a line of muscular, big-boned, "Conestoga" horses from Flemish and English stock. These heavy horses were recognized as the best breed of draft animals during the colonial period and were thus preferred over slow-moving oxen of earlier days.

The number of horses, mules and asses, in 1840 was 4,335,669, twenty years later the horses alone numbered 7,434,688 including those used in cities and towns. By 1920 the United States could boast of a horse and mule population of over 22 million animals. By 1940, trucks and tractors had cut this number almost in half — 6,096,000 farms had 1,610,000 tractors; 935,000 trucks; 4,185,000 automobiles and 13,368,000 horses and mules. Common wisdom of the 1930's was that a farm needed to have at least 75 acres under cultivation before it became economical to replace horses and mules with a tractor.

It is common for a draft horse to have a tractive pulling power of up to 1/8 of his weight and travel at 2.5 miles an hour for 20 miles a day. A typical 1,500-pound draft horse developed one horsepower (33,000 foot pounds per minute) of tractive pull. The average farm horse was not that heavy and developed only about 2/3 horsepower.

Horse-powered treadmills called *horse powers* were developed in the 1830's and probably evolved from the earlier English dog-powered treadmills that were used to turn fireplace spits and butter churns. These treadmills actually multiplied a horse's power output. Working a 1,000-pound horse on an inclined conveyor belt with a slope of 1 to 4 was equal to a pull of 250 pounds, or about 1.33 horsepower. A 1,600 pound horse, treading at about two miles an hour, produced work done at the rate of 2.13 horsepower.

Horses walking on treadmills developed more power than those pulling a *sweep power*, which worked like a merry-go-round, geared to a power take-off. But sweep powers were not as expensive and you could hitch up to 14 horses to the multiple sweep-arms of the largest models. Both sources of cheap four-legged horsepower were utilized to run all kinds of farm equipment, ranging from threshing machines to gristmills, until economical gasoline engines took over in the early 1900's. Some larger farms used steam engines to perform these chores.

The Mighty Ox. Western cattle are descendants of the large European wild oxen that were tamed by Stone Age tribesmen. The smaller Celtic shorthorn was the most important domestic ox of the Stone Age. From these ancient animals came modern cattle, and the North American ox, which, in the more restricted usage is a castrated male used for draft purposes. Many people think of the domestic ox as a specific breed, but actually it is just any cow's son that can't reproduce itself.

Not every bull calf had the right stuff to become a good working animal. His disposition had to be quiet and his intelligence above normal. Certain cows bore calves that made better oxen than others, and they were used for this breeding purpose in addition to giving milk. In the early settlements any cow's male offspring was taken for oxen, but as time progressed certain breeds, such as the sure-footed Devon, with its bright red shading, were highly favored. Much effort was made to make oxen look like matched pairs, even to the point of artificially bending their horns by traction and scraping.

Calves selected to be oxen began their training at the age of two or three months. My father was given a calf to train when he was a young child in rural North Carolina. Dad would fasten a rope around the calf's neck and lead it off into the woods to haul out a small log — he was a miniature Paul Bunyon with his ox, Babe.

By speaking the animal's name and calling "Gee" with a prod to the right, and "Haw" with a gentle shove to the left, a trainer could teach a calf its name and obedience to human commands in a week or two. At the age of three or four months a future ox was fitted with a calf-sized yoke and driven around for an hour or so every day by a patient trainer.

At about six months of age young bulls were forever deprived of their manhood, by a quick, painless procedure that "left them chewing their cuds and flicking their tails as if nothing serious had happened."

After castration, an ox's growth pattern changed, the hindquarters became heavier than normal, and the animals were bigger than bulls of the same age. A full grown ox weighed about a ton. At two years old they were qualified for full-time work as ox cart or light wagon pullers. By their fourth birthday a pair was ready for heavier fieldwork and longer hours. From five to fourteen it was all work and no play, and at fifteen or sixteen years old they were butchered.

A pair of well-broken oxen was worth as much as a pair of horses of the same age. But oxen were cheaper to own in the long run. They ate less food than horses and required half the amount of grain. They could also survive on pasture fare during the occasional idle periods of summer. Like horses, oxen might become overheated and require a restful nap under the shade of a tree. However, oxen were not as sensitive to minor irritations such as falling branches or sliding logs, and they were better adapted to working in snow or deep in the woods.

But in spite of all these advantages, the slow moving ox — traveling about two-thirds as fast as a horse — was almost entirely phased out of work on North American farms by the early 1900's.

HAULING COTTON TO MARKET.

MULES are the crossbred result of mating a male ass, or "jack", with a female horse. George Washington was a great advocate of using mules for farm and plantation work. To that end he improved the quality of the mules produced here by importing the Spanish *asinus* (ass), for cross breeding with American mares. The female mule is seldom fertile and there were no reported cases of fertility in male mules. (Mule colts were generally emasculated at 12 to 18 months of age and very few males survived with their all their private parts intact to demonstrate any fertility they might have possessed.)

The donkey, or domestic ass, is descended from the African and Asiatic races of the wild asinus. Donkeys were domesticated far earlier than horses, and biblical references are made to both asses and mules.

Mules were used extensively for general farm work, but more were sold in the Cotton Belt and adjoining states than in the north or west. A good farm mule measured from 15 to 16 hands tall and weighed about 1,300 pounds. Black was the preferred color because gray mules turned white with age. Style and action were not as important in mules as in horses, but a smart, alert mule with a long, free walking stride and a a snappy trot was highly desired. Farmers who raised horses turned to mule raising when a bigger profit could be made from mules, which was often the case. The market classes of mules were: draft mules, farm mules, sugar mules, cotton mules, mining mules and pack mules.

Advantages of Mules. Mules are more resistant to the effect of hot weather, and are easier to feed since they have fewer digestive problems than horses. The self-feeding method, using corn, oats, hay, etc. works very well with mules, as they are more sensible about overeating. An inexperienced rider or driver is safer on a mule than on a horse. Under most any conditions a mule is far more able to take care of himself than a horse. Mules are cool-headed and less apt to be nervous than horses; consequently they are more inclined to accept hard work, abuse, and poor handling. Mules are less subject to lameness than horses and they also perform better under difficult conditions and on irregular terrain. The overall toughness of mules as work animals accounted for their wide demand and frequently higher value.

Disadvantages. Mules are not as good at pulling heavy loads as horses, and the smaller feet of a mule can be a disadvantage in very soft soil, or on pavement, because they have less surface adhesion. Since mules are strong willed, tough mouthed, and slow in response, there is less pleasure in driving them. You wouldn't want to be caught courting your girlfriend in a "mule and buggy".

NEW HAVEN AGRICULTURAL WAREHOUSE AND SEED STORE

HAND TOOLS. As we mentioned earlier, the first settlers arrived here with very few agricultural tools. Fortunately there were skilled craftsmen among them who could duplicate basic Old World farm implements using native hardwoods and scraps of imported iron.

From these humble beginnings evolved the wide selection of hand tools and agricultural implements pictured in the circa 1860 engraving above. Many of these tools were manufactured in New England and were great improvements over earlier imported and blacksmith-made examples.

If you were to send a frontiersman into the wilderness with his choice of a single tool, he would probably choose an axe. You can shave with an axe, you can kill or butcher an animal with an axe; you can also clear the land, build a house, prepare fuel and protect your family with this versatile tool.

Samuel W. Collins, a Hartford, Connecticut merchant, became convinced that America was ready for mass-produced axes in 1826, and started making them at a small stone building near his home. In 1828 he moved operations to an area near the village of Canton, now named Collinsville. There, using water-powered trip hammers, eight employees began producing sixty-four axes a day. By 1868 the business had become a stock company with capital of over a million dollars, employing six hundred men who manufactured three thousand tools a day. In the 1860's Collins' factory teamed up with an Illinois inventor by the name of F. F. Smith and manufactured 100,000 of Smith's revolutionary forty-pound, cast steel plows during the next ten years.

The first axes were found objects, hand-held pieces of stone, or other hard materials. Flaking a chunk of flint to a sharp edge and tying it to a handle was the next step in the evolution of the axe. Then, about six or seven thousand years ago, someone discovered that they could hammer-harden a chunk of native copper into a useable axe; or better yet, cast an even tougher blade out of bronze, using an air-blown campfire to melt the natural occurring ore.

Around 2,500 B.C., the Sumerians began to produce swords, axes, and other implements in bronze. The Hittites discovered how to forge and temper iron in 1,400 BC and managed to keep the process a secret from the rest of the world for the next three hundred years. By 850 B.C. the wandering Celtic tribes of Central Europe had begun to settle down in villages. The largest of these settlements could easily support a full-time blacksmith, and from a charcoal-fed forge the village smithy produced a variety of iron beads, axe heads, swords and primitive plowshares.

Cast iron did not become commercially useful before the Industrial Revolution and wrought iron remained the favorite material for making axe heads well into the 18th Century. Most *trade* and *felling* axes of the period were not much more than a bow-shaped piece of wrought iron, folded over an anvil and hammer-welded together at one end to form a cutting edge. After 1744, a steel strip was commonly inserted between the two red-hot halves of the cutting edge during this welding process.

There were two basic kinds of axes, *felling* and *hewing*. To chop down, or *fell* a tree, one needed a long-handled, lightweight tool with a cutting edge ground on both sides of the blade. To shape, or *hew* a log into a beam, one needed a short-handled, broad-headed axe with a slightly curved blade, sharpened on one side only, a *broad axe*. Within these two basic categories fell a wide variety of sizes,

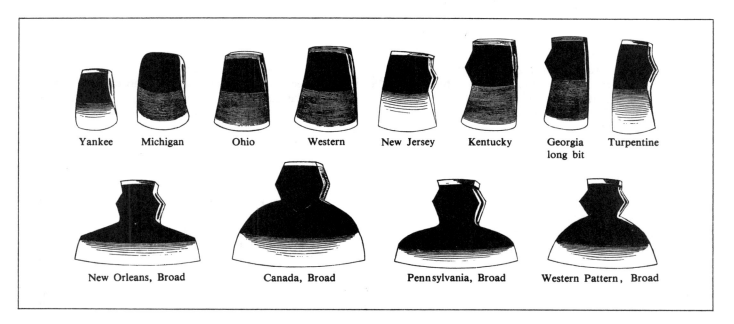

shapes and regional variations. Some favorite patterns were: Yankee, Michigan, Ohio, Western, Kentucky, Georgia long bit, Pennsylvania Broad, Western Broad, and so on; often hammered out by a local blacksmith. (The axes illustrated above are factory-made examples from the 1860's.)

Dibble: a pointed stick, or forked limb, used to prepare soil for seed dropping. An 1803 text said: "The implements for dibbling of beans, nuts and potatoes are: (a) The common dibble, about 18 inches long, with a handle like a spade and an iron shod point. (b) The long dibble, 3 1/2 feet in length, for potatoes. (c) The bean dibble, which is made from a plank of oak with a handle affixed to the middle and wooden pegs, spaced at a proper length from each other, to poke holes in the ground for bean planting. The youngest children may be taught to follow this instrument and drop a bean in every hole thus made."

Drawknives are the ancient shaping tool of most all woodworking trades. This two-handled knife blade, 1 1/2 inches wide and 6 to 15 inches long, is pulled toward the user to shape and shave everything from axe handles, to barrel staves, ox yokes, and wheel spokes. A handy farmer could duplicate almost any wooden implement or broken furniture part with his drawknife.

Flail: a simple wooden club hinged with a piece of leather to a long staff which is used to thresh (beat the heads off) oats, beans, corn and wheat. Each type of grain required a different sized flail. After threshing, the straw was raked away from the loose grain on the threshing floor, and the grain and chaff were shoveled into a tray or winnowing basket. When there was enough wind the mixture was tossed up in the air and the chaff was blown away..

Hand rakes appear in several engravings of 16th century garden scenes, and literature of the period also mentions their occasional employment as weapons.

This incident involving a common hand rake appeared in an early New England record book: "Levi Hubbard, a hot-tempered member of the Mass. General Court, nicknamed 'The Smasher', was ostracized from that body and fled to what is now Rhode Island. There he sold a wooden rake to a savage, who used it in wet sand to dig clams. Naturally the rake soon came apart and the Indian appeared at Hubbard's door, demanding satisfaction. After a heated battle of grunts and words, Levi smacked the Indian in the belly with a stone hammer that he had been holding in his hand, and knocked the savage backward into a berry patch. The red man's squaw, who had been standing at his side, retaliated by pitching her lord's tomahawk at Levi. Her aim was perfect and that was the last of the hot-headed Mr. Hubbard, whose skull was smashed in because of the errant use of a rake."

During the early 1800's most New England and New York towns supported a rake shop, usually attached to a water-powered lumber mill. The shops also turned out handles for other tools as part of their trade. Hickory and ash were the preferred woods. Hickory was limber and tough enough to bend for bows and braces without steaming, and handles made of ash lasted indefinitely. The use of ash for tool handles dates back to the first century A D , when it was used by Romans living in England.

All kinds of rakes were produced at rake shops. There were windrow rakes with angled heads and projecting teeth on both sides for sweeping in either direction. And there were bull rakes for wet hay, charcoal rakes with six inch teeth, and corn rakes with four inch tines.

The first mass-produced rakes were twelve-tooth hay sweeping implements with six-foot handles fastened to their heads by bent wood bows or dowels. English-style hay rakes were usually made of lightweight willow, some with replaceable threaded wooden teeth. A swath rake of the late 1700's was about two yards wide "with iron teeth and a bearer in the middle to which a man fixes himself with a belt. When he has gathered as much grain as his rake will hold, he raises it up and begins again."

By the early 1900's, improved horse-drawn hay rakes had cut into the demand for wooden hand rakes, however, production continued well into the 1940's. After the war, cheap bamboo and metal garden rakes flooded the market.

HAYFORKS, barley forks, manure forks, etc., were first secured from nature. Finding a six-foot long limb with a two-pronged fork was not difficult. Three, four, and six-prong forks were made by splitting the end of a limb and riveting butternut wedges between each prong. In the early 1800's metal tips were added to the wooden prongs. Twin-pronged, wrought iron, hayforks were often hammered out by a local blacksmith.

"The Farmer's Guide" of 1824 mentions metal forks: "In gathering potatoes, it is said that one man can throw out of a hill, with a four pronged metal fork, as many as five or six farm hands can pick up and throw." By 1840, several varieties of cast steel forks were on the market.

Hay knives, hay spades, and hay saws were not used in harvesting hay, they were made for cutting hay in the mow, or sawing off a chunk from a tightly compressed hay stack. Hay knives appear in farm literature of the 1830's and in trade catalogs well into the 20th century.

Hoes had blacksmith-made heads of wrought iron until 1823, when cast-iron hoe factories were first established in Philadelphia. By 1838 several factories had begun production of steel tools. In the 1600's hoes came in only three sizes: small, narrow and large. By the middle 1700's a wide variety of hoes was available. The Virginia Gazette advertised the following styles of hoes for sale: Bristol, broad, fluke, garden, grubbing, harrow, hilling, trowel and weeding. An 1803 text mentions various types of hand-hoes: The prong hoe (with two or three points). The Dutch hoe (operated by pushing). The narrow spud hoe, the common hoe (with a blade from 3 to 12 inches wide), and the Portuguese hoe (with a short handle and a heavy conical shaped iron). Cotton hoes had iron heads large enough to fry pancakes (hoe cakes) on.

Hammers were late arriving in the Colonies. Wrought iron nails were chiefly handmade prior to 1800, and were so valuable that people sometimes burned down houses to get them. Early Americans fastened boards together with dowels, or nails made from native hardwoods. These wooden *trenails* were driven with hardwood hammers or mallets. Claw hammers were not widely used until after the Civil War; shingling hatchets and the flat polls on axe heads served instead. The iron or steel claw hammer of modern times is a Roman invention dating from the time of Christ. Many of today's specialized hammerhead designs came to us from the Middle Ages.

Mattock: a digging and grubbing tool shaped like a pick, with an adze-like hoe on one end of its head and an axe blade on the other. American Indians made them out of the shoulder blades of large animals. Colonists made them from wrought iron and used them to hoe the hardest ground, to cut roots, and to dig ditches and wells.

Saws were highly advanced tools even in the days of the Egyptian tomb builders. Early Roman saws had the "set" teeth of today's crosscut version. By the late 1600's wooden framed saws, with their slender blades, began to be replaced with wider-bladed saws of hammer-hardened steel that did not require the support of a stretcher frame. Water and wind-powered saw mills, producing an up and down action, first appeared in France during the 12th century. New York state and New England had wind and water powered saw mills in the 1630's, but a new saw mill built in Limehouse, Great Britain, was destroyed by a mob of unemployed sawyers in the year 1760

The conversion of timber into shingles, firewood, or dimensioned lumber was an important part of many farmers' income. Saws, driven by horse power -- and later, steam, kerosene, or gasoline -- made this task easier.

Circular saws were introduced in the 1840's that whirred through even the largest logs at 2,400 revolutions per minute, but old traditions did not die quickly; hardware catalogs of the 1880's offered circular saw blades side-by-side with the pit-saw variety of a century earlier.

Shovels for apples and grain were first hand carved from maple, poplar, or basswood; those made for turning the soil were reinforced with iron plates. One of the first implement factories in the United States was a shovel manufacturer established in Pittsburgh shortly after 1800. By 1822 a shovel factory located at Easton, Massachusetts, was turning out thirty thousand tools a year. Iron and steel shovels, spades and forks were available in a wide variety of sizes and shapes after 1830.

Sickles and *Scythes*. These ancient reaping tools date back to Old Testament times and several were probably included with the first boat load of pilgrims. The Romans were mowing hay with *two-handled scythes* a century before the birth of Christ. Somehow these early Roman scythes disappeared during the dark ages and had to be reinvented 500 years later.

Egyptian slaves are pictured on tomb walls, circa 1500 B.C., using short-handled *sickles*, and *reaping hooks* made of flint or bronze.

Grass and wheat were best reaped with different tools. Grass for hay could be mowed and raked "willy-nilly" and tumbled about to dry uniformly. A scythe served nicely for this purpose. Wheat required a more orderly swath for the binders to handle, and a less violent mowing action, so that ripe grain would not fall from the stalks. The smaller short-handled sickle filled these requirements.

The time came, however, for the common two-handled scythe to be modified for use in wheat fields. Cradle-like fingers were added by Flemish farmers in the 16th century, to catch the wheat and lay it in small heaps called *gavels*, for following workers to gather and tie into sheaves.

The American cradle, with five long wooden fingers and a serpentine-shaped handle, appeared in estate inventories as early as 1745. It is recorded that George Washington was hiring "cradlers" for work on his plantation in the year 1766. (Some historians have stated that the use of the grain cradle did not begin in the United States until the early 1800's.)

As late as 1850, grass hooks, sickles, and slightly larger reaping hooks were still used by some farmers who preferred these lighter, short-handled tools because almost any member of a family could wield them accurately. Mowing with the heavier 5-foot sythe required strength, skill, and endurance.

On the following pages are engravings of gardening and horticultural tools from the 1875 catalog of B. K. Bliss & Sons. Not every farmer needed these specialized tools, but it is interesting to note how closely these implements resemble those of today.

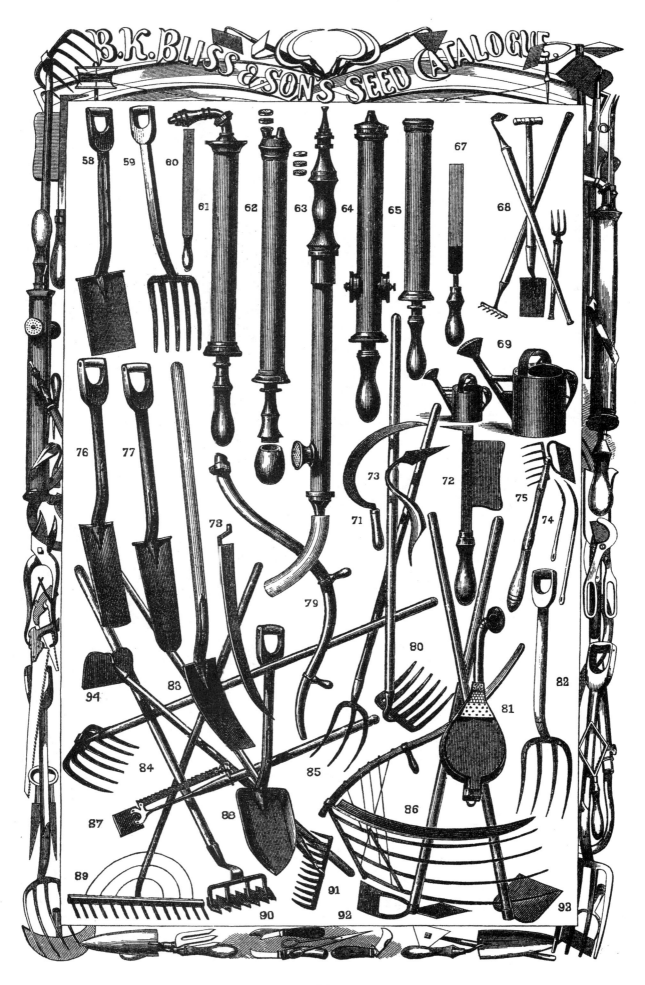

Garden and Horticultural Implements.

From the best English and American Manufacturers. See Illustrations

No.	Item	Price
1	**Pruning Knife**, with Saw, (Saynor's),	$2 00
2	**Pruning Knives.**—(Saynor's,) $1.50; No. 3, $1.75; No. 4, $1.50; No. 5, $1.50; No. 6, $1.00; No. 7, $1.75; No. 8, $1.75; No. 9, $1.50; No. 10, $2.00; No. 11, $1.25; No. 12, $1.25; No. 13, $1.25; No. 14, $1.00; No. 25, $1.00.	
15	**Budding Knives.**—(Saynor's,) No. 15, $2.75; No. 16, $2.00; No. 17, $2.00; No. 18, $1.50; No. 19, $1.50; No. 20, $2.00; No. 21, $2.00; No. 22, $1.25; No. 23, $1.50; No. 24, $2.50.	
	Pruning and Budding Knives of similar patterns, from other manufact'rs,	$0 75 to 1 50
26	**Border or Grass Shears**, 8-inch, $3.50; 9-inch, $4.00; 10-inch, $4.50.	
27	**Border Shears**, with wheel, 8-inch, $4.00; 9-inch, $4.50; 10-inch, $5.00.	
28	**Branch or Lopping Pruning Shears**, three sizes, $3.00, $4.00, and $5.00.	
29	**Hedge and Garden Shears**, 5½ in. (ladies,) $2.50; 8 in. $3.00; 8½ in. $3.25; 9 in. $3.50; 10 in. $4.00; 12 in. $5.00; notched, 25 cents extra.	
30	**Garden Bill Hooks**, for pruning with one hand,	2 00 to 3 00
31	**Spring Grass Shears**, for edging,	1 50
32	**Sheep Shears**, for edgings,	1 50
33	**French Pruning Shears**, with springs, various sizes, styles and finish,	1 75 to 5 50
	Similar pattern of American manufacture,	1 50 to 2 50
34	**Bow Slide Pruning Shears**, 7-inch,	4 50
35	**Pruning Scissors**, with bows, three sizes, $1.00, $1.50, $2.00.	
36	**Grape Scissors**, 6-inch, $1.00; 7-inch, $1.25.	
37	**Propagating Scissors**,	1 25
38	**Scotch Scythe Stones**, 35 cents each; $3.50 per dozen.	
39	**Bayonet Hoe**, without handle,	75
40	**Grass Plot Edging Knives**, cast steel, (Saynor's,) 8-inch, $2.25; 9-inch, $2.50; 10-inch, $2.75.	
41	**Dutch or Scuffle Hoes**, (Saynor's,) 4-inch, 50 cents; 5-inch, 60 cents; 6-inch, 75 cents; 7-inch, 80 cents; 8-inch, 90 cents; 9-inch, $1.00; 10-inch, $1.10.	
42	**English Transplanting Trowels**, blued steel, 6-inch, $1.25; 7-inch, $1.50; 8-inch, $1.75.	
	Similar pattern of American manufacture, 6-inch, 35 cents; 7-inch, 40 cents; 8-inch, 50 cents.	
43	**Triangular Hoes**, used also for Tree Scrapers, 5-inch, 50 cents; 6-inch, 60 cents; 7-inch, 75 cents.	
44	**Noyes' Garden Weeder**,	40
45	**Garden Reels**, with stakes, English, 8-inch, $1.50; 10-inch, $2.00.	
	Similar patterns of American manufacture,	75 to 1 25
46	**Moore's Improved Grafting Knife**,	1 25
47	**English Lawn Rakes**, 16-inch, $3.50; 20-inch, $4.00; 24-inch, $5.00.	
48	**Ladies' Blue Weeding Forks**, English,	60
	Similar pattern of American manufacture,	40 to 50
49	**Claw Hatchets**,	1 25
50	**Pruning Saws**, 14-inch, $1.50; 16-inch, $1.75; 18-inch, $2.00; 20-inch, $2.50.	
51	**Comstock's Weeding Hook**,	50
52	**English Lawn Scythes**,	1 50 to 2 00
53	**English Turnip Hoes**, 6-inch,	50
54	**Asparagus Knife**,	1 00
55	**Excelsior Weeding Hook**,	30
56	**Garden Harrows**,	1 00
57	**Milton Hatchet**, stag handle,	4 00
58	**Ames' Cast Steel Spades**,	2 00
59, 82	**Spading and Manure Forks**, cast steel,	1 75 to 2 50
60	**Rifle for Sharpening Scythes**,	25
61	**Brass Syringe**, $10.00; No. 62, $9.00; No. 64, $9.00; No. 65, $5.00.	
63	**Fountain Pump**, brass, with three feet of hose,	10 00
67	**Asparagus Cutter**,	1 00
68	**Ladies' and Children's Garden Sets**, (4 pieces,) according to size and finish, $2.00, $3.00, and $4.00.	
69	**Tin Water Pots**, painted green, from 2 to 16 qts.,	50 to 3 00
71	**Grass Hooks or Sickles**, English, three sizes, 75 cents, $1.00, and $1.25.	
72	**Grafting Chisel**,	1 00
73	**Vernon Hoe**, three sizes, $1.00, $1.25, and $1.50.	
74	**Sacking Needles**,	20 to 25
75	**Ladies' Floral Rake and Hoe**,	25
76	**Post Hole Spade**, Ames' Cast Steel,	2 25
77	**Draining Spade**, cast steel,	2 25
78	**Scythes of various patterns and manufactures**,	1 25 to 2 00
79	**Scythe Snaths of various patterns**,	75 to 1 50
80	**Cast Steel Potato Hooks or Prong Hoes**,	1 00
81	**Sulphur Bellows**, for preventing mildew, see page 167,	2 50
83	**Spades and Shovels**, long handles, Ames' and others,	2 00
84	**Hexamer's Prong Hoe**, see page 164,	1 50
85	**Hay or Manure Forks**,	50 to 1 50
86	**Grain Cradles**,	3 50 to 5 50
87	**Pruning Saw and Chisel combined**,	3 00
88	**Ames' Round Point Shovel**,	1 75
89	**Wooden Rakes**, of various patterns and sizes,	50 to 75
90	**Allen's Weeding Hoe**, different sizes,	1 25 to 1 75
91	**Steel Garden Rakes**, 6 teeth, 75 cents; 8 teeth, 80 cents; 10 teeth, 90 cents; 12 teeth, $1.00; 14 teeth, $1.20; 16 teeth, $1.50.	
92	**Weeding Hoe**, old pattern,	1 00
93–94	**Garden Hoes**, cast steel, various sizes and patterns,	75 to 1 00

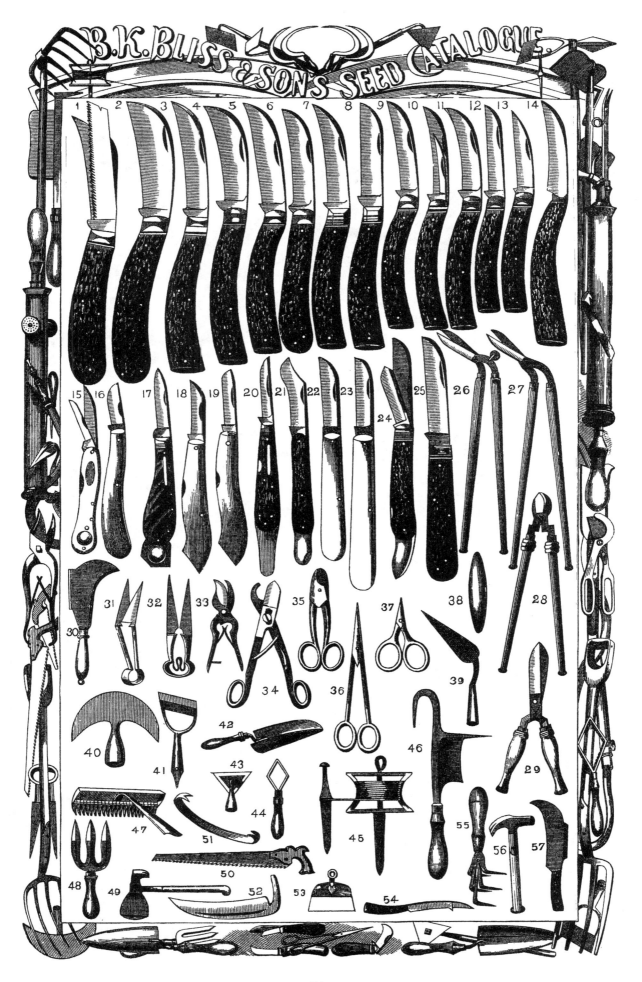

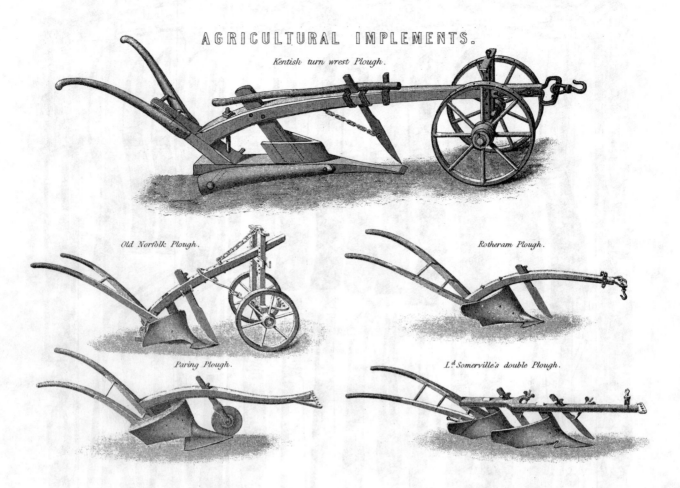

A selection of British plows in use from 1730 to 1800.

THE PLOW was the most important tillage implement a farmer owned. Virgin ground had to be opened up and turned over completely to bury all the weeds and native grass. On established farms the soil needed to be turned, broken up and mixed with air, manure and vegetation in order to decay and form valuable plant food.

Over a span of twenty years several different styles of plows evolved in America: the *lister*, the *middlebreaker*, the *disk plow* and the *regular moldboard* plow were all used in turning, stirring and pulverizing the soil in preparation for planting. Each plow bottom was named for a specific soil condition: the prairie breaker, stubble bottom, black land-bottom, slat moldboard, brush plow, sandy land plow, cotton plow, ditching plow, contractor's plow, etc. There were plows for tough sod, old ground, sticky soil, gumbo and "buck-shot" soils. You name it.... by 1850 there was a plow designed to tame it.

The first plows used in this country were clumsy wooden affairs, not unlike those of early Egyptian or Roman times. A pointed clump of wrought iron or a piece of hardwood served as the plowshare. These primitive tools did little more than dig a groove in the soil that was a little bit deeper than the roots of the surrounding weeds and brush. They were commonly referred to as "hog plows", because of their propensity to root in and out of the ground.

Southern planters found that a simple shovel-shaped wrought iron plowshare, fastened to a wooden beam, worked best to prepare the sandy soil for their traditional crops of cotton and tobacco.

A history book published in 1870, states: "As early as 1617, some ploughs were set to work on the Virginia plantation, but in that year the governor complained that the colony did suffer for the want of skillful husbandmen who knew how to plow: 'We have as good ground as any man can desire, and about forty bulls and oxen; but we need men to labor, iron for ploughs, and harness for the cattle. Some thirty or forty acres we had sown with one plough, but the crop stood so long before it was reaped that most of it was shaken off and the rest spoiled by cattle and rats in the barn.'

The author further states: "It is recorded that in 1637 there were but thirty-seven ploughs in the colony of Massachusetts Bay. Twelve years after the landing of the pilgrims, the farmers about Boston still owned no ploughs, and were compelled to break up the bushes and prepare for cultivation with their hands, and with rude hoes or mattocks"

"It was the custom in that part of the country, even to a much later period, for anyone owning a plough to go about and prepare the ground for others in the surrounding territory. A town often paid a bounty to anyone who would buy and keep a plough in repair for this purpose."

MEANWHILE, in Massachusetts, a much heavier plow evolved that consisted of two sturdy planks joined in a "V" shape. Several oxen were required to pull this massive contraption. The plank that threw up the furrow was called the *mouldboard* (later spelled moldboard). The plank, which ran along the trench, shoring up the earth, was called the *landside*. When farmers figured out that friction was slowing things down they began to fasten scraps of sheet iron to the moldboard and share, to resist the clinging soil and prolong the plow's working life.

The most popular plow design made by local blacksmiths of the early 1800's was called the Carey plow. It was a primitive looking implement made almost entirely of wooden parts, all doweled together behind its huge wooden moldboard and wrought iron share. Even though it was a fairly light implement, strong arms were needed to hold it in the ground and it required twice the pulling power of a factory-made English plow. Many farmers continued to use old Carey plows long after mass-produced implements became available. The old Carey plow's simple wooden construction enabled field hands to make on-the-spot repairs.

In England, France, Belgium and Germany, plows had already become highly refined tools. The Dutch plow was imported into Yorkshire in the 1730's, and served as a model for early English plows such as the circa 1800, "Old Yorkshire" style illustrated above. This was the first notable improvement over the early Roman plow, which was described by Virgil as being made of two pieces of wood meeting at an acute angle and plated over with pieces of iron. The early Dutch plow had most of the characteristics of later plows, including a curved moldboard, beam and two handles.

The Royal Society of Arts, founded in 1754, sponsored plowing contests all over the British isles; even using a dynamometer in 1784 trials, to determine the proper draught required for peak performance. P. P. Howard, of England, and James Small, of Scotland, were early improvers of the plow. Howard's plows were provided with a bridle, or clevis for regulating the depth and width of the furrow and his factory remained in business for more than a hundred years.

Robert Ransome patented a tempering process for cast-iron plowshares in 1785, and developed a self-sharpening chilled-iron share in 1803. Then in 1808, he started making a knockdown version with interchangeable bolt-on parts.

By 1840 Ransom's Ipswich, England, factory was offering eighty-six different models to a worldwide market. A sizable number of English plows were sold in America, where implement factories were just becoming established.

Charles Newbold, a New Jersey farmer, invented the first *American* cast-iron plow in 1797. (James Small had pioneered the use of cast iron plowshares in Scotland, in 1740). Newbold spent several years, and his life savings of $30,000 perfecting and promoting his metal plow, which except for the beam and handles, was a solid piece of cast-iron. However, he was forced to abandon the entire project after local farmers became convinced that "Cast-iron poisons the land and stimulates the growth of weeds." They also pointed out the fact that "When such a plow is fractured the whole must be replaced."

Thomas Jefferson made public, in 1798, his most popular invention; a plow designed on scientific principles, featuring a curved, metal-shod, wooden moldboard with a flat, triangular-shaped front blade. The straight-bladed share cut under the sod, and the gentle curve of the moldboard raised the soil and turned it completely over. Later, Jefferson had the entire moldboard cast in iron and in a test run, pulled by two small horses, it easily plowed a furrow nine inches wide and six inches deep.

A Philadelphia writer of the year 1803 related that: "Ploughs without wheels are suited for stony, uneven soils; the turn-wrist plough is adapted to hilly land, its mouldboard turning to suffer the plough to also work upon its return and lay a furrow while traveling down the hill. The advantage of one or two wheels is that they keep the share at a uniform depth. Some ploughs have two shares, the first skims the turf and lays it in the bottom of the former furrow, and the other brings up fresh mould in order to cover the turf.

THE EVOLUTION OF PLOW DESIGN

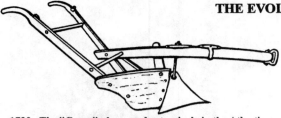

1790. The "Carey" plow, used extensively in the Atlantic states until the 1820's. It varied in design according to the maker, but consisted of a wrought iron share with a landside and moldboard plated over with scraps of sheet iron.

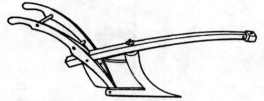

1797. Charles Newbold invents the first American cast-iron plowshare. But farmers reject the rust-prone metal as being "poisonous to the soil".

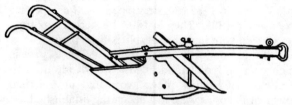

1798. Thomas Jefferson introduces his superior light weight plow with a scientifically designed metal-shod moldboard, knife-style coulter, and triangular shaped share.

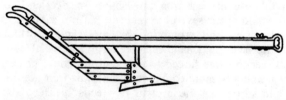

1833. John Lane uses discarded steel saw blades to sheath his streamlined design. Lane pioneered the development of "soft center" steel plowshares in the 1860's.

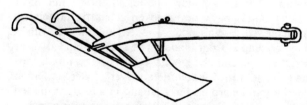

1837. John Deere fashions the first steel moldboard and share from a giant circular saw-blade. Mass-production begins in 1842.

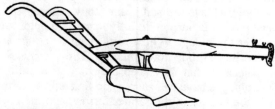

1873. James Oliver patents the first American plow with replaceable parts and it becomes the model for future plow design in the United States.

IN 1819, JETHRO WOOD improved upon Thomas Jefferson's moldboard design and added interchangeable bolt-on iron parts. Some authorities say that Wood gave the American plow its proper shape. *The Farmer's Guide* of 1824 had nothing but nice words to say about the new metal plows: "Woods' New York cast-iron plough can be moved with power one quarter less, and it will do one quarter more work than any other plough in use in the state. One hundred and twenty acres were ploughed by four of these ploughs in one season; and the whole expense of keeping them sharp and fit for use did not exceed 56 cents each. The blacksmith's bill for ordinary wood and wrought iron ploughs would have been about $5 each."

The new plow was a big seller in the Eastern United States, where farmers frequently broke their plowshares in a constant battle with the rocky soil. However, Mr. Wood spent most of his profits fighting numerous patent infringers and died in relative poverty. William Seward said of him: "No man has benefited his country pecuniarily more than Jethro Wood, and no man has been as inadequately rewarded."

By the 1830's, cast-iron had become widely used in plowshare making. We quote from Thomas Fessenden's 1842 edition of *The Complete Farmer*: "The cast-iron plough is now the most generally used among farmers, and considered decidedly the best. Among the different ploughs made of this metal, Charles Howard's, of Hingham, Mass., stands unrivaled. There has been no plough presented since 1832 which has been considered better than Howard's. They have been exhibited at all the cattle shows and have won the most prizes in ploughing contests. Recent improvements made by him have resulted in making the mouldboard longer than usual, and swelling the breast of the share, which makes every part bear equally. The new plough runs more true and steady, is always free from carrying forward any earth, and wears perfectly bright." Meanwhile, out west, near Chicago – where soil conditions were vastly different than in New England – the mud and molasses muck was sticking to iron plow blades in great gooey chunks that required constant cleaning and slowed things down to a snail's pace. In the 1840's, even a good day's plowing did not turn over more than an acre of ground. Self-scouring steel seemed to be the long sought answer to this persistent problem.

The first steel plow made in America, circa 1833, was actually a steel *plated* wooden moldboard, sheathed in three horizontal rows with discarded crosscut saw blades; a reinforced triangular scrap served as the cutting share. Its inventor, John Lane, and other pioneer steel plow makers, had to postpone production seven years, until 1840, before domestic rolled steel became available from mills in the Pittsburgh area.

John Deere, a twenty-nine year old Illinois blacksmith, noticed that nothing seemed to stick to the giant circular steel saw blades in use at the local Grand Detour lumberyard. In the summer of 1837 he talked the lumberyard owner out of a discarded Sheffield steel blade and took it home to his forge for tempering, cutting and reshaping. The final planishing was done with a wooden hammer, over a hardwood log, in order not to mar the moldboard's perfectly smooth surface.

John Deere's "New Deal" sulky plow of 1887

JOHN DEERE'S first plow sold for ten dollars and he could barely keep up with the resulting demand for his new steel-bladed sod-buster. By 1842 Deere was turning out twenty-five plows a week, utilizing a crew of blacksmith recruits working ten hours a day for $25 a month (including their room and board).

Production continued to increase at a rapid pace and blacksmith-made plows had become a rarity by 1858, the year John Deere's factory shipped 13,000 of his "singing" plows to western farmers. Deere had plenty of competition for the farmer's dollar. William Parlin had established a plow factory in 1842 at Canton, Illinois, which, by 1900 was one of the country's largest. International Harvester bought out Parlin's "P & O" line of plows and implements in 1919.

Among other noteworthy plows were Joel Norse's "Eagle" brand, circa 1842, and John Lane's "Soft Center" steel implement of the 1860's. Lane's concept of using a harder, sometimes brittle steel for the outer surface and forming a much softer resilient steel for the inner core of the plowshare, was eventually adopted by most other makers. It is recorded that at least two hundred different steel-bladed plows were on display at a Western plow show in 1859.

James Oliver developed a method of hardening cast-iron that made it very competitive with the more expensive steel plows. His "chilled iron" process involved super-cooling hot cast iron with water, resulting in a stronger, lighter plow, which became a best-seller of the 1870's.

Another success story is that of **F.F. Smith**, who had begun hammering out sheet steel plows in the late 1850's in a tiny Illinois prairie town. Mr. Smith read of a new casting process developed in England, whereby church bells were being cast in solid steel, rather than bronze or cast-iron. Smith figured he could use a similar process to make cast steel plows, and in 1860 he struck a deal with Sam Collins, the Connecticut axe manufacturer.

After signing a partnership agreement, Smith moved to Collinsville where he practically lived in Sam Collins' factory, working night and day to perfect his casting process. It took Smith a year to solve the problem of cooling the fragile castings without any resulting cracks. After the casting process was refined Collins' company began to manufacture of Smith's revolutionary forty-pound plow and sold more than 100,000 of them over the next ten years.

SULKY PLOWS, with two or three wheels and an added seat, allowed foot-weary plowmen to ride upon — rather than walk behind — the implement; but they were not an immediate market success. The first heavy wood-framed prototypes of the 1850's met with skepticism by farmers, who feared that the added draught would be too much of a burden on their already overworked horses.

Eventually it became obvious that the new metal-framed riding plows were more easily handled than walking plows, and that the sulky increased a farmer's production while it lightened his task. Walking plows, however, did not disappear. Deere and Company continued making them well into the 1940's.

F. S. Davenport patented a popular two-wheeled sulky plow in 1864 and another inventor, Robert Newton, added the large disk-style coulter which became an industry standard.

The Greatest ···AND··· Most Popular Plow on Earth!

The Friend of the Horse and The Delight of the Farmer.

More Flying Dutchman Jr.'s

in the field to-day than any other sulky plow manufactured and the army of Juniors swelling into innumerable numbers.

The Popular Verdict:
**THE LIGHTEST DRAFT,
THE EASIEST OPERATED, AND
THE BEST WORK DONE.**

The frame being made entirely of steel, makes it the

STRONGEST and MOST DURABLE

and the shapes of plows used add to its perfection. Can be used with Sod Breaker, Old Ground or Scotch Clipper plows to cut all practical widths of furrows. Any boy who can manage a team can do the work of a man.

ESPECIALLY MADE FOR TURNING SQUARE CORNERS

Others try to immitate, but none equal the **Flying Dutchman Jr.** It is made with both Right and Left Hand. It was the first three-wheel Sulky ever made. Do not be deceived by buying anything else. If your dealer can't supply you, address,

MOLINE PLOW CO., Moline, Ill.

BY THE YEAR 1875 there were more than three dozen brands of sulky plows available to American farmers. One of the best sellers — based on the Gilpin Moore patent of 1875 — was manufactured by John Deere.

James Oliver's Chilled Plow Works also produced a successful two-bottom sulky plow — the first to use a wheel for a landside — patented by W. L. Casaday in 1876.

The Moline Plow Company, of Moline, Illinois, was quick to capitalize on the new trend, and by the late 1880's had captured a lion's share of the market with their Flying Dutchman Jr., the first three-wheel sulky ever made.

As stated in the 1888 advertisement pictured above, its biggest claim to fame was the ability of an operator to turn corners at a right angle without stopping his horses forward motion. The new lightweight steel frame could be fitted with a variety of plow bottoms and even a boy could operate it.

The name *"sulky plow"* was used for all wheeled-plows, but applied more specifically to single plows, while the name *"gang"* was given to double-bottom or larger plows. It is interesting to note that the rear and front wheels, which are set at an angle, are called *furrow wheels* because they have to carry the side pressure from turning a furrow slice. The largest wheel traveling upon the unplowed land is called the *land wheel*.

The two-bottom gang plow doubled the number of acres that one man could plow in a day and by the 1920's was the most widely used implement in the country.

The three-bottom gang plow appealed to operators in the big-farm districts, especially in the Northwest, where even greater savings in time and man power were of utmost importance.

Tractor plows differed from horse-drawn plows in that they had a tractor hitch, but no seat. Like horse-drawn plows, tractor-drawn plows were built in a variety of styles, having from one to four bottoms. Some farmers converted their horse-drawn implements to motorized usage, but most bought larger plows that were suited to the power of their tractors.

Disk plows were used on dry, hard land that moldboard plows could not penetrate, and in sticky, waxy, and gumbo soils. They were also used to advantage in very loose ground and in stony or rooty land. A plow consisting of three disks, cutting very narrow strips, was among the first disk types patented; the Cravath brothers, of Bloomington, Illinois were its inventors.

Modern farming methods have made traditional plows almost obsolete. Today, the idea is to disturb the soil as little as possible before planting, and the disk harrow has widely replaced the moldboard plow as a primary tillage implement.

B. F. AVERY & SONS, 1915

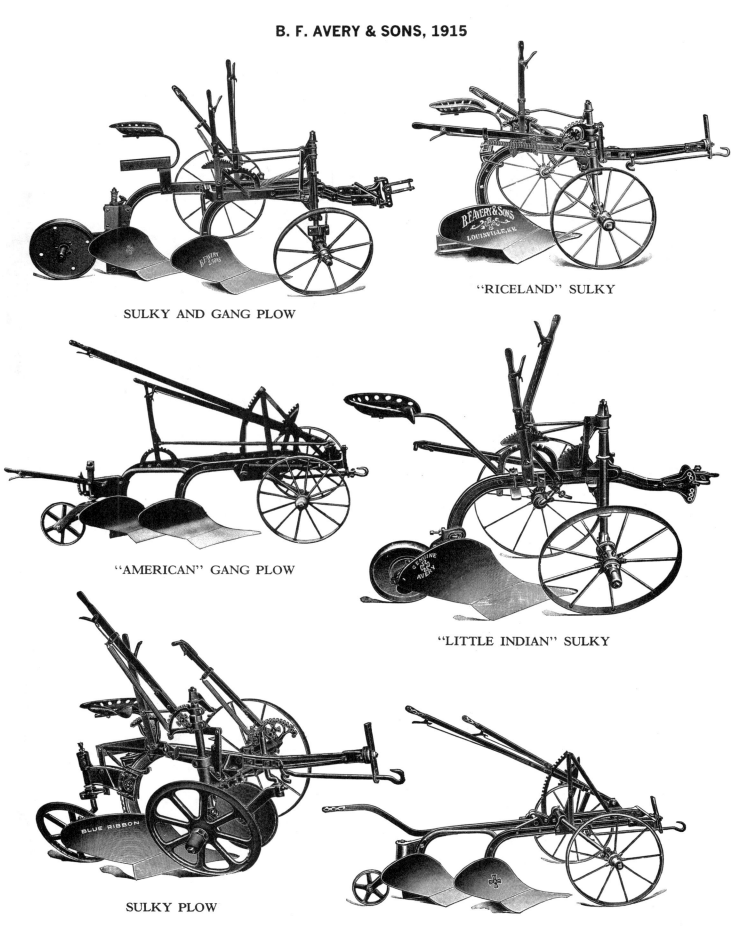

Farmhands drive their disk harrows over freshly plowed ground in this 1902 photo.

THE HARROW followed the plow over the ground, pulverizing dirt clods, packing and leveling the soil, and leaving a surface mulch ready for sowing. Harrows were also used to uproot weeds, aerate the soil, and cover seeds. Primitive harrows were fashioned from twiggy tree branches or thorny bushes, often weighted down with stones, or a human rider. These *brush harrows* were still used to cover seeds in some localities until the late 1880's.

The Roman harrow consisted of a plank with iron spikes protruding through its bottom. Another early type of harrow consisted of a forked limb with spikes in each arm. Later came the traditional rectangular and triangular wood-framed versions with pointed hardwood teeth. Steel-tipped iron teeth replaced the wooden variety in the 1830's. These two harrows were generally superseded by the "zigzag" style harrow, patented by Armstrong in 1839; after which came the steel lever harrow, with a riding attachment.

The metal *disk harrow* was an American invention of 1847. The first version had only one disk, which sufficed until 1854, when a second disk was added. By the 1880's self-cleaning models with a dozen disks were available.

Flexible, deeper penetrating *spring-tooth* harrows came on the market in 1877 and were big sellers in rock-strewn areas of the Central and Eastern United States. They were also used to eradicate quack grass and other obnoxious weeds. The traditional spike-toothed finishing harrow remained in use well into the 20th century.

An agriculture textbook of 1909 encouraged frequent harrowing: "Fall-plowed land is usually left without any other working until spring. If heavy soil is fall-plowed and too finely pulverized, it is likely to run together. Spring-plowed land should be dragged with a smoothing harrow (the traditional spike-toothed style), or otherwise stirred before the clods become too dry to crumble readily. The drier the soil the more frequently this should be done. Under normal conditions, the harrowing should be performed the same day as the plowing. If the weather is very dry, particularly in semi-arid regions, it may be necessary to harrow within a few hours after plowing. One may stop in the middle of each half-day for this purpose. Usually the land should be harrowed with the smoothing harrow two to four times before planting. Sometimes it may be better to use a disk harrow."

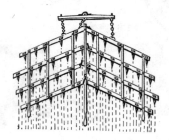
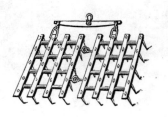

1884
"Acme" pulverizing harrow, leveler and clod crusher.

1888
Janesville "Budlong" disk harrow.

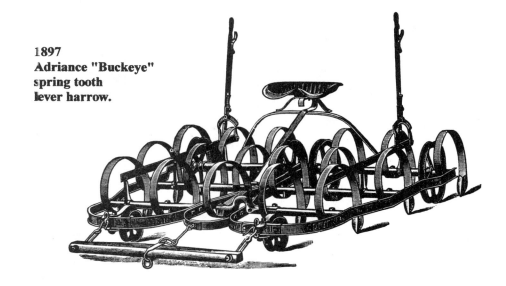

1897
Adriance "Buckeye" spring tooth lever harrow.

A circa 1918 Two-way "Success" Sulky Plow at work turning over freshly-harrowed soil. Some farmers used harrows before and after plowing in order to obtain an ideal seed bed.

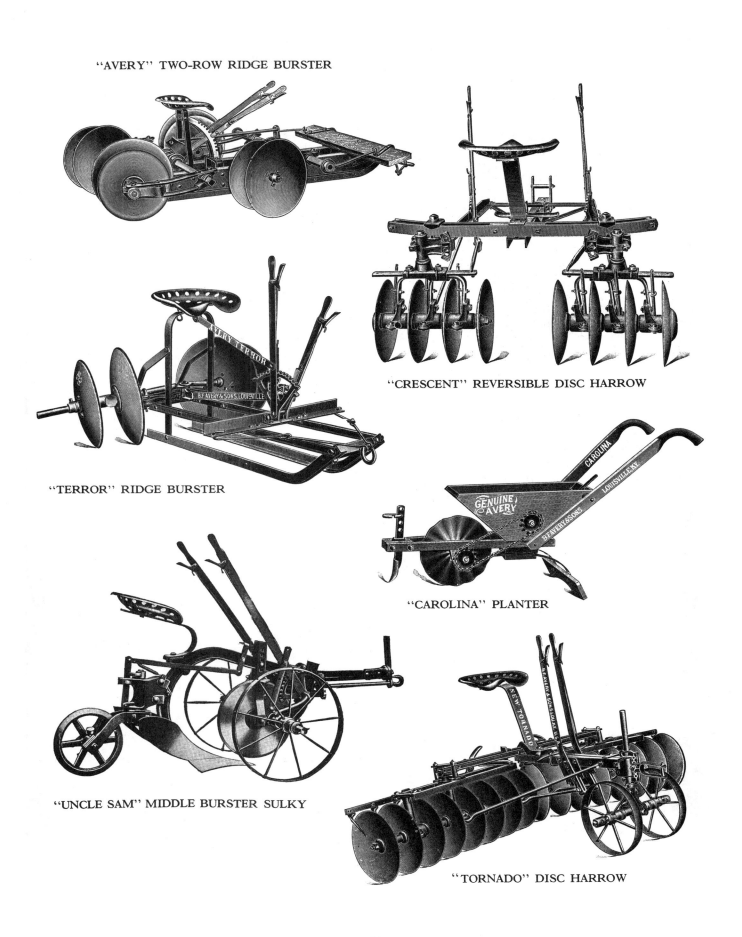

1750 engraving from Diderot's Encyclopedia of Trade and Industry. Fig. 1. A French farmer guides his two-wheeled plow of the type invented by English agriculturist, Jethro Tull. Fig. 4. A woman pushes a wheel-activated seed-dropping machine along a newly opened furrow. This seeder was invented by Abbe' Soumille, an 18th century French priest. Fig. 5. In the background a worker broadcasts grain by hand. Fig. 6. Another worker covers the seeds with a horse-drawn spike-toothed harrow. Fig. 7. A heavy roller breaks up clods and firms the soil after plowing.

SEED DRILLS, grain drills, or seed-sowers, have been around for more than a thousand years. The ancient Babylonian seed dropper consisted of a funnel-and-tube arrangement, attached just behind the triangular blade of a shallow-digging plow. An attendant walked along beside the plow, dropping seeds and fertilizer in the funnel as he went.

An Englishman, named Jethro Tull, is credited with the invention of the first practical seed-dropping machine in 1730. However, the practice of dropping seeds by hand into soil prepared with a pointed stick, called a *dibble*, continued well into the 19th century.

In 1731 Jethro Tull published a work entitled *Horse Hoe Husbandry*, in which he argued that grain should not be broadcast, but should be drilled (dropped in a depression) in rows and cultivated. Tull designed a machine that would drill three rows of turnips or wheat at a time. He used a coulter as a furrow opener and planted his seed at three different depths; saying, that if one seed failed, the others would come up later and be successful. Jethro Tull spent a lifetime developing a successful line of drills, horse-hoes, and cultivators, but he died poor, as did most inventors of his day.

There were a few early attempts at developing an American seed drill, but wheat, barley, oats, rye, and grass seed were usually tossed about by hand, as evenly as possible until well after mechanical *broadcasting distributors* appeared in the 1850's.

By the 1860's broadcast seeders were available in many styles, ranging from fiddlebow, to wheelbarrow-style and wagon mounted models. The *knapsack seeder* — a canvas bag mounted on an iron frame, with a hand cranked revolving distributing wheel — was popular from 1860 to 1940. It was suspended by a strap from the operator's neck and held in front by a waist strap. At a common walking gait you could sow from four to eight acres an hour. *End-gate seeders*, which operated from a chain sprocket on a wagon wheel, were also broadcast-type seeders, but like hand-cranked broadcasters, they could be foiled on a windy day.

The *wheelbarrow-style* that appears in trade catalogs of the 1850's - 1920's, seems to be a survivor of an earlier English seeder. In America it was used to sow grass seed by means of a vibrating rod which passed underneath its long, trough-like, horizontal box, causing the seeds to flow out of openings in the bottom.

The fact that uniform planting depth and seed distribution was critical to obtaining a maximum crop yield was obvious. The brothers Moses and Samuel Pennock of Chester County, Pennsylvania, tackled the problem and came up with a much-improved version of the ancient Babylonian grain drill. Their 1841 model had a cylindrical grain hopper instead of a funnel and the amount of seeds dropped down its tubes could be regulated. Individual cylinders could also be thrown in and out of gear while in operation. Moses and Sam Pennock manufactured their own grain drills and enjoyed a virtual monopoly until after the Civil War.

Among the later seed drills were specialized versions of *force feeders* and *slide feeders*, with hoes, shoes, disks, wheels, and tubular prongs that planted everything from alfalfa to beans, beets and cotton. *Slide drills* differed from other types in that a slide was employed to vary the size of the opening through which the seed passed, thus controlling the amount of seed sown.

Force feed drills carried the seed from the hopper into cavities in a seed cylinder, in which the amount was varied by the size of the pockets, or the speed of rotation. The first patent for a force feed drill was issued to the gentlemen N. Foster, G. Jessup, and the brothers H.L. and C.P. Brown, on November 4, 1851. This was the introduction of the term "force feed". The Brown Brothers incorporated the Empire Drill Company and established a factory at Shortsville, New York, in 1854.

Shoe drills came into use about 1885 and had many advantages over the earlier *hoe drill* that simply covered the seeds with dirt, or other surface refuse. The curved shoes, however, were pressed into the ground by independent springs that discouraged clogging with trash.

Disk drills came later. Shaped like a harrow, they had a heel, or shoe, to insert the grain in the bottom of the furrow. Some versions deposited grain through the center of the disk.

Press wheels, which went out of style around 1900, compacted the earth around each seed and caused moisture to be drawn up to it, thus enabling early germination.

We should mention here that each type of drill was developed for a particular crop, or section of the country. In wheat territory, where the ground was not plowed every year, a drill with great penetration was needed. In other sections where the ground was carefully prepared, deep penetration was not important.

During the 1880's and 1890's large-scale farmers used grain drills up to sixteen feet wide. These big machines were pulled by four-horse teams and could plant 20 acres a day.

In the 1930's three principal types of grain drills remained in use: the *fluted feed*, the *double-run feed* and the *fertilizer grain drill* — each one having a different feeding mechanism. The *fertilizer grain drill* completed four operations at once. It pulverized the soil, planted the seed, distributed fertilizer and then covered them both.

All of the above types could be accurately calibrated, but operators were advised to jack up the drill beforehand, over a catch cloth, and revolve the wheel at the same speed it would travel at work. The resulting grain dropped to the canvas and was measured to calculate how much would fall on an acre.

A STANDARD SINGLE-DISK DRILL WITH A PRESS-WHEEL

JONES DOUBLE PLANTER.

A FARMER carrying a sack of grain and using only his hoe could plant about an acre of corn in ten-hours. When hand-held, *lever-operated, corn planters* were first introduced in 1852, the number of acres that could be seeded by one man increased twenty-fold. Hand planters remained in catalogs for decades after corn drills were perfected. They served in small gardens and also to fill in, or "plug" in corn that was missing from an emerging row.

Early *horse-drawn seeders* were the same implements as those used for smaller grains. The first machine invented exclusively for the planting of corn was patented in 1839, by D. S. Rockwell. It was a primitive implement, but had some features of 20th century corn planters. The furrow openers were vertical shovels, and the planter was mounted between a set of front and rear wheels. Kernels of corn were dropped by means of a slide opening beneath the seed box.

Many other patents followed: *check row markers* in 1855 and 1857, and various styles of automatic seed drops thereafter. George W. Brown, of Galesburg, Illinois, secured patents in the 1850's on several features of the corn planter, including the shoe-style furrow opener, the rotary drop, and press-wheels. Brown built his own factory in Galesburg to manufacture his inventions.

During the 1850's some farmers began to plow their fields in a checkerboard pattern in order to more thoroughly cultivate the emerging shoots and kill more weeds than straight one-way plowing had accomplished. Many early attempts were made to develop a practical automatic check-row seed dropping mechanism. Most of them involved the use of chains, knotted rope, or wire to activate the seed dispenser. The first successful automatic *check-row planting machine* was invented by G. D. Haworth in the 1870's. His machine utilized knotted steel wire.

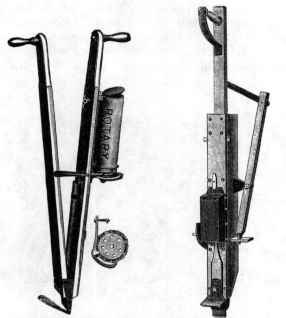

"ROTARY," KEENEY & HARRISON.   "CHAMPION," WINSHIP MFG. CO.

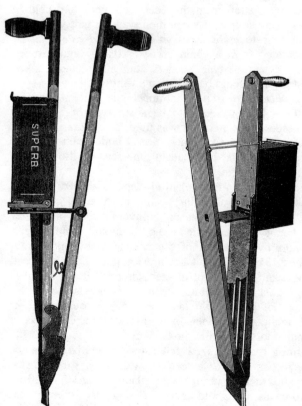

"SUPERB," HORTON MFG. CO.   "ARNOLD," ARNOLD & DENTON.

"LISTING" was a method of planting corn, cotton and wheat in regions with limited rainfall. The *lister* was a combination plow and planter with a V-shaped moldboard that threw the ground in both directions, leaving a deeper trench than conventional planting implements. The first listers were used by the Assyrians, about 3,000 years ago and reinvented by the Italians in 1600.

English manufacturers began making *listing plows* with attached seed hoppers in the early 1800's. An Illinois farmer, C. Atkinson, patented the first American listing plow in 1860.

The following comments regarding listing techniques were made in letters from farmers to the editor of *Farm Implement News* in an 1888 issue:

"Instead of planting corn in traditional ridges it is deposited at the bottom of the listed furrow where the stalks and roots are protected from the hot, dry, prairie winds."

"By planting the seed in the bottom of a furrow, and then filling in the furrow with an attached covering shovel, or disk, the plant's roots grow deep below the surface, where moisture is more plentiful."

"With the weed seed thrown out into the middle of the row the corn is more easily tended, and in the process of working the corn while small, all old stalks, husks and other trash are thrown into the furrow and covered with dirt. There they serve as a moisture-holding mulch during dry weather."

Listed crops were more easily cultivated and kept free from weeds, but a major drawback was that the system would not work well in poorly-drained level areas where frequent rain could collect in the deep furrows and drown the emerging crop. Neither could listing be used in hilly locations where the corn would tend to wash out. In districts where the soil was suitable, and winter wheat was raised on a large scale, listing proved to be the fastest way to prepare a seed bed.

John Deere and Co., offered a *listing plow* and a *two-row listing cultivator* in their 1930 catalog:

"These implements have not only aided in the production of wheat at lower costs, but in some of the more sandy and dry sections of the country have established new methods that save moisture and increase crop yields."

"The listing plow and ridge burster are used in conjunction. The listing plow throws up ridges, forming a surface mulch and moisture trap, and the ridge burster – with its multiple discs – levels the field. Listing is usually done immediately after harvest, and ridge-bursting just before planting."

Single-listing consists of planting and covering the seeds in the bottom of the furrow in one simple operation. Blank-listing is plowing the seed bed with a listing blade without planting, leaving the planting to be done later with a regular corn planter. Double-listing is the practice of blank-listing in the fall to catch the snow and hold the moisture, then splitting out the ridges and single-listing in the spring.

Listed crops required special two-row cultivators with hooded shields to prevent dirt and trash from being thrown into the furrow and smothering the emerging plants

(The top two listers illustrated on this page are 1888 models. The double row lister at the bottom is from a 1920 catalog.)

COMBINED LISTER.

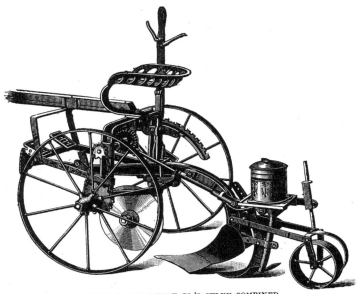
FARLIN & ORENDORFF CO.'S SULKY COMBINED.

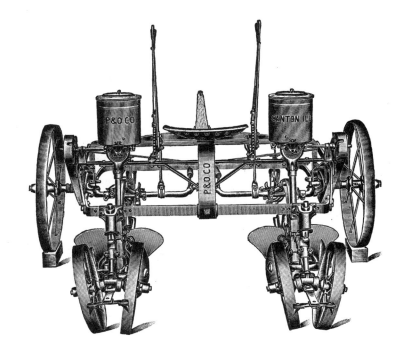

Farm Implement News, 1888

HOOSIER DRILL CO., Richmond, Ind.

THE IMPROVED HOOSIER
TWO-ROW TWO-HORSE CORN DRILL,
WITH COG GEAR.

Strong and durable, furrows the ground, drops and covers two rows at a time, and marks for return rows. Drops one grain in a place, and can be adjusted to deposit them either 12, 16 or 20 inches apart, to suit the requirements of various kinds of soil. It has three adjustments in width, which is an important feature. Light running. Easily handled and a machine of rare merit.

THE HOOSIER CORN DRILL,
WITH COG GEAR.

In every way superior to any other Corn Drill. Dealers can handle it with pleasure as it never fails to give farmers entire satisfaction. Furnished also with detachable fertilizer attachment. Drops one grain in a place, and can be adjusted to deposit them either 8, 12, 16 or 20 inches apart, as may be desired.

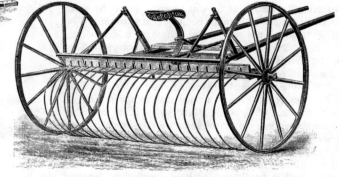

THE NEW HOOSIER HAY RAKE

Is practically a self dump, will rake and bunch the largest windrow. The teeth are made of the best material and are the most approved pattern and style. Elegantly finished. The **Truss Rod** prevents all sagging of the head. We furnish these Rakes with **Combination Pole and Shafts**, thus enabling the farmer to use one or two horses, as may be desired. Can be fully relied on to do its work expeditiously and well.

GEO. W. BROWN & COMPANY
MAKE

THE BEST Planters AND Check Rowers

Write for Catalogue and Prices. Mention this Paper.

ASK FOR THE KNOX SHELLER.

THE BEST Cultivators and Shellers

AGRICULTURAL IMPLEMENTS

B. F. Avery and Sons, 1915

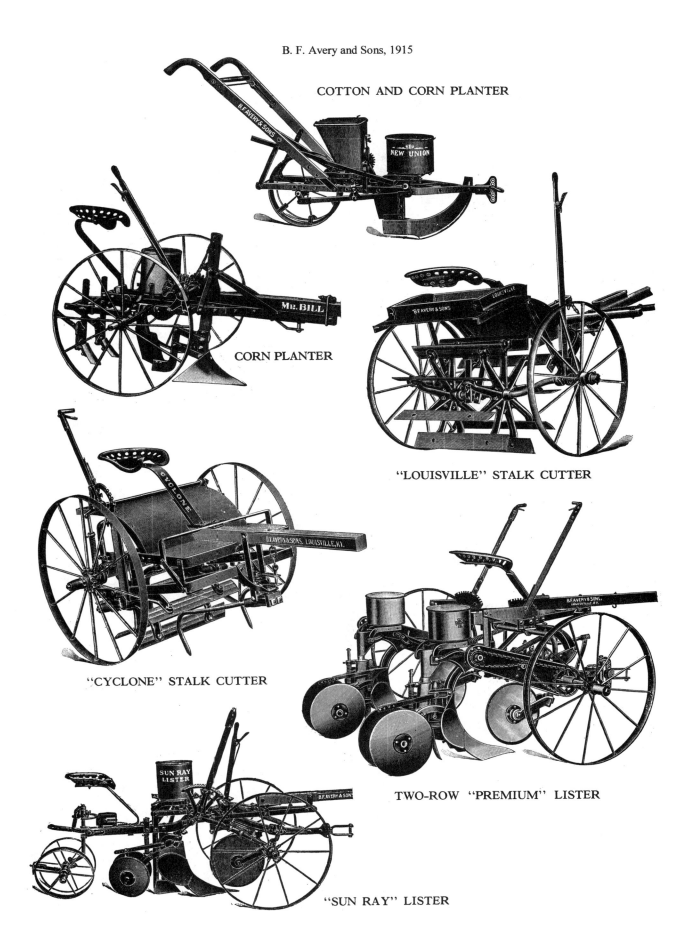

NEW JERSEY Manure Baling Co.

For Descriptive Circulars apply to

JOHN R. BURNETT,
Box 165, Newark, N. J.

Baling Yard, Corner of Oliver St. and New Jersey Rail Road Ave., Newark, N. J.

SOLUBLE PACIFIC GUANO.

YEARLY SALES 50,000 Tons.

We again offer this old established and reliable Fertilizer, which during the past year has fully sustained its high reputation. For Farm Crops of all kinds, Fruit Trees, Gardens, Lawns, and Flowers, it has no superior.

For sale by our agents throughout the United States. Pamphlets containing testimonials and directions forwarded free.

Glidden & Curtis,
General Selling Agents,
Boston, Mass.

THE JOHNSTON HARVESTER CO.,
Manufacturers of
Self Raking Reapers, Combined Machines, and Mowers.
Send for Catalogue. Brockport, N. Y.

MEDAL MACHINES.
New York State Agricultural Works.

First Premiums at all Competitive Trials.

Railway, Chain and Lever Horse Powers, Threshers and Cleaners, Threshers and Shakers, Clover Hullers, Feed Cutters, Wheel Horse Rakes, Horse Pitchforks, Shingle Machines, Straw Preserving Rye Threshers, Portable Steam-Engines, Cider and Wine-Mills and Presses, Dog and Pony Powers, etc., etc.

WHEELER & MELICK CO.,
ALBANY, N. Y.

Send stamp for Circular and report of Centennial trial.

THE FAVORITE POTATO BUG EXTERMINATOR

Is the best and most satisfactory Machine in use for dusting poisons mixed with plaster, etc., on Potato vines.

SAVES TIME, LABOR, and MATERIAL

The detestible Potato Bug it **DESTROYS** Safely, Quickly, Effectually

PRICE $1.50

Liberal terms to agents & dealers
Write for full description and terms. Manufactured by
J. S. EDDY & SONS, Eagle Mills, Rens. Co., N. Y

STEAM ENGINES,
A. B. FARQUHAR, York, Pa.

Cheapest and best for all purposes—simple, strong, and durable. Also *Traction Engines* for common roads.

SAW, GRIST AND RICE MILLS, GINS, PRESSES AND MACHINERY generally. Inquiries promptly answered.

Vertical Engines, with or without wheels, very convenient, economical and complete in every detail, best and cheapest Vertical in the world. Fig. 1 is engine in use. Fig. 2 ready for road.

Send for Catalogue.

The Farquhar Separator (Warranted) Penna. Agricultural Works, York, Pa. Lightest draft, most durable simplest, most economic l and perfect in use. Wastes no grain, cleans it

Steam Engines, Horse Power and Threshers of all kinds. Also, Ploughs, Cultivators, &c., &c.

Send for Illustrated Catalogue.

CHEAPEST FERTILIZER
BECAUSE THE BEST.
PREMIUM BONE

Grows WHEAT and GRASS, equal to manure, at Half the Cost, and lasts twice as long.
Farmers!! Send questions on Postal Card to EXCELSIOR FERTILIZER WORKS, Salem, Ohio. Circulars Free.

THE WESTINGHOUSE
THRESHING MACHINES & HORSE POWERS
Best at Centennial Trial.

Threshers for all sizes of Horse Powers and Steam, Lever Powers for 4 to 10 Horse, and Endless Chain Powers for 2 and 3 Horses—and Engines from 4 to 10 Horse Power. All with late and important improvements. Send for Circular.
G. WESTINGHOUSE & CO., Schenectady, N. Y.

CHAPMAN & VAN WYCK,
(Established 1849),

DEALERS IN PERUVIAN Guano

EXCLUSIVELY,
170 Front Street,
New York.

KEMP'S MANURE SPREADER
PULVERIZER and CART COMBINED.

Greatest Agricultural Invention of the Age! Saves 90 per cent. of labor. Doubles the value of the Manure. Spreads evenly all kinds of manure, broadcast or in drill, in one-tenth time required by hand. Illustrated Catalogues free.
KEMP & BURPEE MF'G CO., Syracuse, N. Y.

THE SILVER & DEMING
Endless Chain Horse Powers,

 for one, two, and three horses, are the best made. They have steel track rods, wrought iron links, and for strength and durability are unsurpassed.

Our Improved Governor regulates the speed perfectly. We also manufacture Drag and Circular Sawing Machines.

Silver & Deming Mfg.

"ECLIPSE"
The Original Self-regulating Solid Wheel.
VICTORIOUS AT WORLD'S FAIRS.
Centennial, '76, Paris, '78, Australia, '80, Atlanta, '81, Chicago Railway Exposition, '83.

Farm Pumping, Grinding Irrigation, Drainage, &c.

Adopted by U. S. Government and all **LEADING RAILROADS.**

Tested 17 years.

TWENTY SIZES. 3-4 to 40 Horse Power.

The Strongest Mill Built.

Also, Standard Feed Grinders, Pumps for House, Farm and R. R. use, and a full assortment of Stock and Reservoir Tanks. Address,
ECLIPSE WIND ENGINE CO., Beloit, Wis.

La Dows Jointed Pulverizing & Smoothing DISC HARROW.

Light, Simple, Durable & Flexible.

Being jointed in the center, is adapted to both smooth and uneven surfaces. Acknowledged the best of the kind, and will pulverize and cover seed better in one operation, than going over twice with others. Made with both Chilled Metal and Cast Steel Discs polished. Send for circular and price list. Manufactured by
WHEELER & MELICK CO.,

MANURE SPREADERS. Now, here is a subject that I can really sink my feet into! During a prolonged period of writer's block in the winter of 1989, I chanced upon a tiny 32-page book about a carpenter who specialized in outhouse construction. *The Specialist*, which was first published in 1929, had a cover blurb that said over a million copies had been sold. I figured that if a million books on outdoor plumbing had been sold during the depths of the Great Depression, perhaps I could duplicate the deceased author's success with my own study of early privies.

To make a long story short, we mortgaged our home and self-published ten thousand copies of *The Vanishing American Outhouse*. It proved to be the biggest challenge of any book I had ever written. We began our research effort by soliciting early privy photos nationwide, from both amateur and professional photographers. Then I hunted down copies of every outhouse book that had ever been published (about 35), and began my own quest for advanced privy lore.

When the delivery truck arrived from our local printer I pulled off the first five boxes of books and began mailing complimentary review copies to dozens of newspaper and magazine editors all over the country.

About two months later we received an illustrated book review in *The New York Times* and over the next decade we managed to sell fifty thousand copies of our self-published coffee-table-sized outhouse book.

Now, back to manure spreaders. As mentioned earlier, the Indians taught the first settlers in America how to grow corn, using fish as a fertilizer, i.e.: "According to the manner of the Indians, we have manured our ground with herrings, or rather shad, which we have in great abundance, and take with ease at our doors."

"You may see in one township a hundred acres together set with these fish, every acre taking a thousand of them; and an acre thus dressed will produce and yield as much corn as three acres without fish."

Later settlers were not uniformly enthusiastic about handling dead fish and cow dung. As a writer of the 1830's observed: "Many farmers allow manure to accumulate and waste away in great heaps, generating effluvia intolerably noisome and perpetually pestilential, without fear of fever or famine, both of which are courted by such conduct. Not only is dung allowed to waste its richness on the tainted air, but straw and other litter is suffered to grow moldy and consume dry rot, both of which might be prevented by covering them with earth."

"Some farmers hang dead cats, dogs, etc., in the forks of fruit trees to improve their yield; but they would be better served to cover the dead animals with five times their bulk in soil, mixed with lime, and allow their decomposition to impregnate the soil with this soluble manure."

The first manure spreaders were homemade affairs, as noted in this early 1800's, *Memoirs of the Pennsylvania Agricultural Society*: "My barn is constructed according to the best models. On the east and west sides are cow stables, containing one hundred and ten well made stalls, ventilated by a number of windows and doors. At the tails of each row of cows there is a drain made of strong planks, so placed to receive all their dung and urine. These several drains are inclined to carry all this liquid manure into a cistern, fifty feet long. This cistern is so placed to receive not only the urine of the stables, but also the liquid matter of the farm-yard. In it there is a pump, by means of which its contents are pumped into a large hogshead, fixed on a pair of wheels and drawn by oxen. To the end of this hogshead is attached a box pierced with holes, into which this fluid manure floats through a faucet, and upon moving is sprinkled over the ground."

The chief source of commercial fish fertilizer during the 1840 - 1860 period was the Long Island Sound. "There, immense schools of 'bony fish' were taken with vertical nets laid by boats and drawn into the shore by great force." One hundred and fifty cart loads was the average catch from one setting of these huge nets. Most of this fertilizer was sold to nurserymen and truck farmers with land close to the big city.

Meanwhile, in the Deep South, planters continued to rapidly deplete their soil by raising tobacco or cotton on the same acreage, year in and year out, without any manure or crop rotation. When the land gave out they just moved a little farther west and did the same thing all over again. Many northern farmers were just as lax and ill informed.

According to deed books of the late 1700's the average northern farm covered from one to two hundred acres; seemingly a large amount of land for one farmer and his children to care for with primitive agricultural tools. But only a very small part was under cultivation at any given time. In eastern Massachusetts a hundred acre farm might have only six acres in crops and ten or twelve acres in meadow and pasture land. The rest, if it had been cleared, lay fallow until the cultivated acres were exhausted of nutrients, at which time the crops were moved elsewhere.

When their fields finally quit producing a profit, most farmers either cleared new land, started raising livestock on a larger scale, or sought a fresh start on the western frontier.

The average farm in eastern Massachusetts at the beginning of the 19th century, had an inventory of one or two horses, two to four oxen, fifteen head of cattle, a few sheep, and a dozen pigs that for the most part ran wild in the woods. A noted historian summed up the status quo:

"The farmer's family could live in modest comfort from this kind of careless agriculture, and there he stopped. He had little money income since he had little or no surplus. He raised no surplus because, generally, he had no market. He had no market because there were no large cities to consume what he might wish to sell, and, if there had been such demand, he could not economically have hauled his produce through bogs and over trackless wooded hills to reach it."

There were exceptions, of course. In the year 1800, farmers in New York and Virginia, who had land convenient to water transportation, shipped 78,000 quarters of wheat to England, who was at war with France. A year later America supplied almost four times as much to the British Isles, where there had been a great crop failure. In 1806 Napoleon conquered Prussia and our farmers again received record prices for grain exports. Meanwhile, the average American farmer remained incredibly conservative. He yoked the same number of oxen to his plow, planted the same number of acres "in the old of the moon", and generally did everything just as his father and grandfather before him had done.

Sears, Roebuck and Co., 1908

$49.85 THE BONANZA FORCE FEED MANURE SPREADER

COMPLETE WITHOUT TRUCK

OUTFIT CONSISTS OF MANURE SPREADER BOX COMPLETE WITH FORCE FEED ATTACHMENT WITHOUT TRUCK. IF TRUCK IS WANTED ALSO WE CAN FURNISH AS QUOTED BELOW

THIRTY DAYS' FREE TRIAL ON YOUR FARM

This machine is guaranteed to be superior to any other Manure Spreader on the market, regardless of price. You can use it thirty days, during which time you can compare it with any other make of machine, regardless of price, and put it to every test. If at the end of that time you are not convinced that it is easier running, that it spreads the manure more evenly and tears it apart better, if you do not find that it can be adjusted more quickly, that you can vary the number of loads to the acre more accurately; in fact, if you do not find that it is a superior machine in every respect to every other machine on the market, regardless of name or make and that you have made a saving of from $35.00 to $75.00, you can return the spreader to us and we will refund the money paid for it, including the freight charges.

FOUR FEEDS—EVERY OPERATION CONTROLLED FROM THE SEAT.

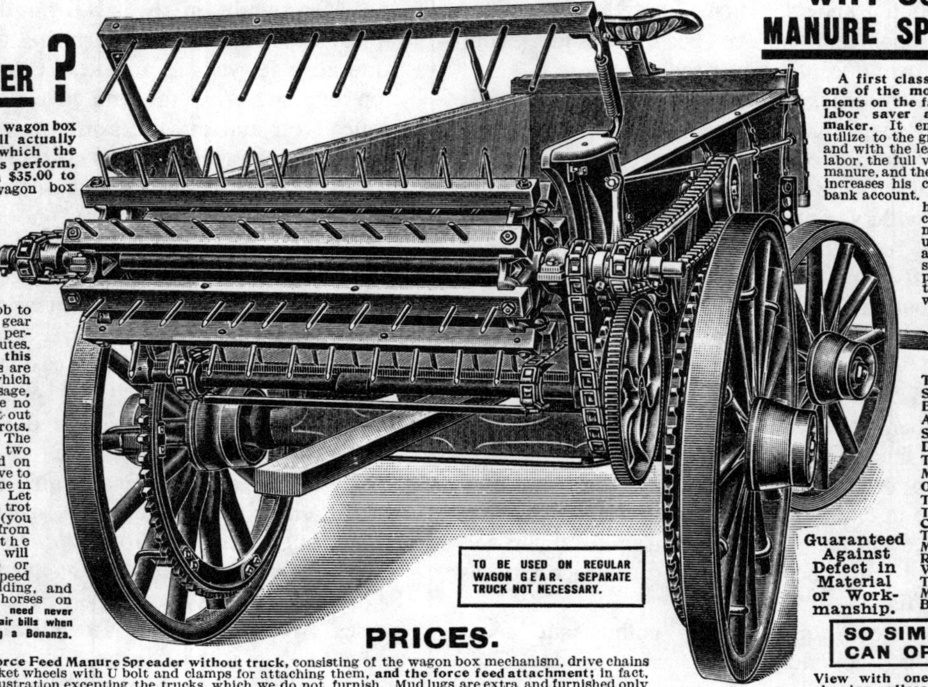

WHY USE A BONANZA SPREADER?

Because it is the only successful wagon box spreader on the market that will actually perform better, all operations which the highest priced manure spreaders perform, and because it will cost you from $35.00 to $75.00 less. It is the only wagon box spreader made which is perfectly adjustable, and which will operate successfully on all widths of farm wagon gears. Our manufacturers are the only ones who have successfully solved this problem, and they have done it in such a surprisingly simple manner that you will find it the easiest kind of a job to place the Bonanza on your wagon gear and to adjust it so that it will be in perfect running order in a very few minutes. It is next to impossible to break this machine, because all working parts are made of malleable iron and steel, which with anything like reasonable usage, should last a lifetime. There are no complicated gears, cogs, etc., to get out of order. The steel apron never rots. There is no strain on the gear. The machine is so light of draft that two horses can easily handle a full load on any kind of soil, and you do not have to stop your horses to set the machine in motion or to change the speed. Let your horses go on a walk or on a trot and work either lever you wish (you can do it from the seat), the machine will start, stop or change its speed at your bidding, and with your horses on the go. You need never worry about repair bills when you are running a Bonanza.

MUD LUG.

TO BE USED ON REGULAR WAGON GEAR. SEPARATE TRUCK NOT NECESSARY.

WHY USE A MANURE SPREADER?

A first class manure spreader is one of the most profitable implements on the farm; it is a wonderful labor saver and a great money maker. It enables the farmer to utilize to the greatest possible extent and with the least possible amount of labor, the full value of his barnyard manure, and thereby enriches his land, increases his crop and adds to his bank account. Scientific experiments have demonstrated that common barnyard manure is the most valuable of all fertilizers, also that it should be spread while it is comparatively fresh, so that the valuable quality which would otherwise be lost in the barnyard will enter into the soil which is to produce the crops.

THE WORK OF SPREADING MANURE BY HAND IS SLOW AND COSTLY, TO SAY NOTHING OF ITS DISAGREEABLE FEATURES. A GOOD MANURE SPREADER OVERCOMES ALL THIS ALMOST ENTIRELY, AND BECAUSE IT SPREADS THE MANURE MUCH MORE EVENLY, THE RESULTS IN CROPS WILL BE FAR BETTER THAN IF THE MANURE IS SPREAD BY HAND.

Guaranteed Against Defect in Material or Workmanship

SO SIMPLE A BOY CAN OPERATE IT

View with one sideboard removed showing endless steel apron, also how the force feed attachment works.

PRICES.

PRICES are for the complete Force Feed Manure Spreader without truck, consisting of the wagon box mechanism, drive chains and the two large sprocket wheels with U bolt and clamps for attaching them, **and the force feed attachment**; in fact, everything we show in the large illustration excepting the trucks, which we do not furnish. Mud lugs are extra and furnished only when ordered and proper price allowed. Complete instructions for putting together and operating accompany each machine. Shipped knocked down from factory in Central Iowa.

No. 11L975 50-Bushel Bonanza Manure Spreader Box for use on narrow track standard gears measuring 38 inches between bolster stakes. Weight, 650 pounds. Price....**$49.85**
No. 11L976 60-Bushel Bonanza Manure Spreader Box for use on wide track standard gears measuring 42 inches between bolster stakes. Weight, 730 pounds. Price..........**$54.85**
No. 11L978 Set of Eighteen Mud Lugs. Weight, 9 pounds. Price **1.65**

There are some who wish to purchase a manure spreader complete, that is, box, truck and all attachments ready to hitch to. We therefore quote below prices on our Bonanza Force Feed Manure Spreader, complete with high grade 3x9-inch cast skein truck, 3x⅜-inch tires, wheels 3 feet 4 inches front by 3 feet 8 inches rear, without neckyoke or whiffletrees. If you buy this manure spreader complete with truck you will secure an entire outfit superior to any complete manure spreading machine on the market at a saving to you of $35.00 to $75.00.

PRICES ON BONANZA MANURE SPREADER BOXES, COMPLETE WITH TRUCK.

No. 11L979 50-Bushel Bonanza Manure Spreader, complete, with 3x9-inch cast skein truck, 3x⅜-inch tires; wheels 3 feet 4 inches by 3 feet 8 inches; narrow track only; 38 inches between bolster stakes. Weight, 1,300 pounds. Price...........................**$82.60**
No. 11L980 60-Bushel Bonanza Manure Spreader, complete, with 3x9-inch cast skein truck, 3x⅜-inch tires; wheels 3 feet 4 inches by 3 feet 8 inches; wide track only; 42 inches between bolster stakes. Weight, 1,380 pounds. Price.........................**$87.60**

DESCRIPTION

THE BONANZA FORCE FEED MANURE SPREADER is built in a strictly first class and workmanlike manner, all parts are made of the best material the market affords, which is selected with a view of providing material best suited for the requirements of each individual part. **THE WAGON BOX** of both the 50-bushel and 60-bushel machines is 10 feet long and 15 inches deep inside; the width inside of the 50-bushel machine is 36 inches, and of the 60-bushel machine 40 inches; the bottom is made of matched lumber and is so strongly braced that it is practically watertight; the sides are smooth dressed and made of the best possible class of material, guaranteed to give perfect service and not to warp. Seat is made of pressed steel, mounted on a spring standard, as shown in the illustration. **PAINTING**—The Bonanza is finished in the best possible style, being painted red with the sideboards neatly striped with black, and varnished. The beater bars are painted to match the box. All iron and steel parts are painted black. The box is stenciled "The Bonanza Spreader" on each side. Our name does not appear anywhere. **THE ENDLESS STEEL APRON** is made of heavy angle steel bars, placed about 12 inches apart and riveted at each end to heavy malleable link chain belts which travel over four sprockets, two at each end of the spreader. **THE BEATER WHEEL** has heavy cast heads, to which the bars are bolted. The bars are cross riveted at intervals of about 9 inches, making it impossible for them to warp or split. The shear pointed teeth are ⅜-inch round, solid high carbon steel and will not bend. No two travel in a line; they throw the manure outward and away from the center. **THE EQUALIZING CLUTCH** overcomes all variation in speed of the rear wagon wheels when making turns or working on other than a straight pull, always insures uniform speed at both ends of the beater wheel, no matter whether the machine is driven straight or when turned in either direction. The mechanism is simplicity itself. All parts are iron and steel and designed to insure perfect action and long service. The large sprocket wheels are perfectly formed and attached to the wagon wheels by means of strong U bolts and clamp

wheel will remain stationary. The amount of manure to be spread per acre is regulated by the lever at the right of the seat, which regulates the action of the dog on the large ratchet wheel and governs the speed at which the conveyor will travel. You can make the machine feed at any one of its four speeds and spread any quantity, ranging from five to thirty loads per acre. **THE IDLER SHEAVES** make it impossible for chains to jump sprockets, even when out of line. These sheaves are mounted on movable studs which can be adjusted to meet the requirements of different widths of gears. This provides great adjustability, and as the small sprockets on both ends of the beater shaft can also be moved outward or inward, it is a very simple matter to adjust the machine to suit any width of wagon gear, and the chain will run true on the sprockets at all times. **THE FORCE FEED ATTACHMENT** shown above, is a most valuable device, consisting of a detachable push board which when set at the head of the machine, engages with the apron, and follows the load, forcing everything before it in a steady stream, clearing the box of every atom. It trips itself before it reaches the beater wheel, and the apron travels free of it. It is very desirable for use in distributing powdered fertilizer, potash, lime, sand, gravel, etc. We furnish the Force Feed Attachment without extra charge. MUD LUGS are listed under No. 11L978 and are desirable when the machine is used in ground that is muddy or covered with ice. Ordinarily these are not required, but will be found useful at some time or other. THE MUD LUGS are extras and furnished only when ordered and proper price allowed.

YOU SAVE IN FREIGHT when you buy the Bonanza Manure Spreader; this is another way in which we save you money. Other styles of manure spreaders weigh from two to four times as much as the Bonanza, and consequently will cost you from two to four times as much for freight charges. This extra weight does not benefit you in any way, because no machine can be stronger than the Bonanza, and again, a heavy machine requires more horses to pull it and cannot be successfully used on

FARMERS, dairymen, and truck-gardeners living on the fringes of large cities had more incentive to improve their methods. Profits brought the desire for increased knowledge and production. The farmland near these lucrative markets grew more and more valuable and was cut into smaller and smaller pieces that required intensive cultivation.

But it was not until 1840, when Liebig came out with his book, *Organic Chemistry*, that literate farmers were made aware of the importance of various forms of fertilizer other than dried fish, animal carcasses, and barnyard manure. "To manure an acre of land with just forty pounds of bone dust, is sufficient to supply *three crops* of wheat, clover, potatoes, or turnips with a very beneficial amount of phosphates."

Immediately after Liebig's announcement experiments were made that substantiated his claim and factories were established in England that required the importation of shiploads of bones from North and South America. Subsequent scientific investigations confirmed that a sea fowl excrement, called *guano*, was the most valuable fertilizer ever discovered for the nourishment of plants. From 1851 to 1868, the United States imported 823,400 tons of guano from various islands frequented by sea birds, off the coast of Peru.

Previous to these discoveries American farmers and horticulturists had confined their fertilizer usage to animal and plant manure, or simple substitutes such as ashes, salt, soot, lime and plaster of paris (gypsum). One man and a boy could cover up to forty acres a day with plaster, using a slow moving wagon pulled by two oxen.

The use of commercial fertilizers rapidly increased, but as of 1900 little was used west of the Mississippi River. In 1879, progressive farmers in the United States spent $29 million for fertilizers, in 1889, $38 million and in 1899, $55 million. They bought everything from sodium nitrate to ammonium sulfate, potassium nitrate, dried blood, slaughter house waste, hoof meal, steamed bone, dried fish, linseed-oil meal, cottonseed meal and phosphate rock. But this was a drop in the bucket compared to their use of the two cheapest and best fertilizers available at the time, barnyard manure and "green-manure". Green crops that were grown to be plowed under as green-manure every five years, were rye, buckwheat, cowpeas, and crimson clover. All of these plants contain beneficial amounts of Nitrogen and other chemicals.

An experiment conducted in 1908 on Cornell University's model farm illustrated the value of barnyard manure on a crop of timothy hay. Using no manure at all produced 2,230 pounds of hay per acre. Applying ten tons of manure yielded 4,350 pounds of hay an acre, and using the maximum recommended application of twenty tons of manure per acre resulted in a bonanza hay harvest of 7,420 pounds.

Believe it or not, pig dung was one of the most valuable manures a farmer could sink his shovel into. A 500-pound porker was good for seven tons of excrement a year – packed full of nitrogen, phosphoric acid and potash, worth a total of $40.30 in 1910 dollars. A 1,200-pound horse would produce about eleven tons of "road apples" annually, which together with his absorbent straw bedding would make about fourteen tons of manure, worth $45.00. The difference between the value per pound, of hog leavings and horse apples, was the quality and richness of the food they ate. Hay-fed horses just did not produce the same high quality of fertilizer that pigs – who would eat anything you gave them – excreted. By the same token, chicken guano was even more potent. A flock of 250 hens, weighing four pounds each, was calculated to generate 4.3 tons of chicken droppings, worth $68.15. Grape growers and orchard owners could not get enough of the stuff.

Manure spreaders saved time and money. Instead of shoveling cow pies out of the barn and into a pile, and then later moving it onto a wagon, and then again pitching it out of the wagon onto the field in irregular, wasteful heaps – a *manure spreader* could be filled daily as the waste accumulated, and when full, driven into the field where a regulated amount could be quickly and evenly spread. Under the old method – where piles were left in the barnyard for months at a time – more than half the nutritive value was lost from prolonged exposure to wind, rain and sun. When manure was spread on fields by hand it was not pulverized and the resulting large clumps often did more harm than good.

According to Davidson and Chase *(Farm Machinery, 1908)*: "The first attempts at the development of a machine for automatically spreading fertilizer were contemporaneous with a machine for planting or seeding. In 1830 the Krause brothers, of Pennsylvania, patented a machine for distributing plaster or dry fertilizer. It consisted of a cart with a sloping bottom and a transverse opening with a roller mounted underneath. The roller was driven by a belt from one of the wheel hubs."

"The first apron machine was invented by J. K. Holland, of North Carolina, in 1850. Its endless apron was attached to the rear end board of a cart and passed over a bed of rollers and around a geared-down shaft. The apron drew the fertilizer to the front and caused it to drop over little by little."

"J. H. Stevens, of New York, produced the first wagon-type spreader in 1865. His machine had an apron that was driven rearward by gears to discharge its load. Vibrating forks fed the manure to fingers extending to each side."

Several other patents were issued for wagon-mounted spreaders in the 1870's, and by the early 1900's most farmers who used more than a hundred wagon-loads of manure a year, owned a commercial spreader that did the work of five men.

By the 1920's these implements had evolved into two basic styles: *apron spreaders* and *tight-bottom* spreaders. An *apron spreader* had an endless conveyor belt, made up of slats, the width of the wagon bed, which carried manure to a beater mounted on the rear axle. A revolving rake aided in holding the upper part of the load until it was shredded by the beater. The operator could control the speed of the conveyor by a feed lever, from the driver's seat.

Tight-bottom spreaders worked on the same principle, but had three beaters instead of one. The upper and main beaters shredded the manure and a spiral beater, at the very end of the chain, spread it evenly over the entire width of the wagon tracks. From five to twenty wagon loads of manure could be spread over each acre, depending on the adjustment of the feed lever.

CULTIVATORS have been around, in one form or another, since the beginnings of agriculture. Modern cultivators evolved from the crude hoes pulled by oxen in ancient times. The destruction of weeds is their primary function but proper cultivation also creates a moisture-saving surface mulch and aerates the soil. An 1803 text describes two styles in use at the time: "The Nottingham cultivator has in common with the plow, a beam, wheels, and handles. Instead of a share there are two parallel 4-foot bars into which 3 inch teeth, 2 feet long, are affixed about 12 inches apart. It is drawn by four horses and can plough six or seven acres a day, in sandy land. The Cook cultivator consists of a diagonal beam, with from 3 to 7 shares of different sizes. It is guided laterally by two handles which force it into the ground."

Most cultivators of the 1830's were heavy-timbered, plow-like implements, with no wheels, levers, or other depth adjustments. But they were considered as being among the most advanced agricultural tools of their day.

Until the late 1850's country blacksmiths accounted for the largest portion of double shovel-type cultivators used on most American farms. It was not a difficult implement to make and farmers needed two or three different kinds; a push-style cultivator for the vegetable garden, a one-row model for standing corn, and a two-row cultivator with a three or four-horse hitch for farms of more than twenty acres. One-row cultivators were built in both walking and riding versions.

From the 1860's through the 1920's there were hundreds of small cultivator manufacturers in the United States and each region favored its own design. There were cotton cultivators for the southern market, disk cultivators for moving large quantities of dirt to or from corn, listing cultivators, surface cultivators, walking and riding styles, etc.

For those farmers reluctant to buy a new sulky style implement – not wanting to further burden their horses – creative manufacturers came up with such enticing names as "The Joy Rider", "Jack Rabbit", "Lucky Boy" and "Success".

Tractor-drawn cultivators came into popular usage in the late 1920's. John Deere's 1928 *Farm Machinery Manual* summed up the many advantages of a tractor cultivator:

"Its adaptability to cultivating row crops is one of the chief reasons for the fast growing popularity of the general-purpose type of tractor. Field experience has proven that tractor cultivators effect great savings in time and labor.

The steady speed of a tractor cannot be matched by horse; especially on hot days. Owners of tractor cultivators find they can put in more hours per day in the field; their cultivating capacity is not limited to the endurance of animal power. In rush seasons, cultivating can be finished sooner and the time saved can be used in taking care of other crops. When the weather is unsettled, a tractor cultivator can quickly take advantage of favorable conditions."

Horse-drawn cultivators continued to be widely manufactured well into the 20th century. The poverty of the Great Depression and the shortages of World War II postponed the purchase of a tractor cultivator for many of our nation's six million farm families until the late 1940's.

(The cultivators at right are from a 1913 John Deere Catalog.)

JOHN DEERE RIDING AND WALKING CULTIVATORS.

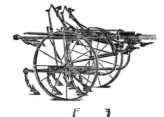

RA Cultivator.

Only one operation is required to raise rigs and balance cultivator instead of three, as on most cultivators. This means a saving of time. Depth of cultivation—each rig independently—is regulated at top of each mast. A variation in width of 40 to 54 inches is possible.

SA Cultivator.

Is a combination riding and walking cultivator with horse lift. Operator simply releases a thumb latch, the horses raise rigs and balance cultivator. A lever for each rig regulates depth of plowing. Adjustable spread arch makes it adaptable to any crop. It is light in draft.

KA Cultivator.

Adaptable for work in corn, potatoes, tobacco, cotton, beans, cabbage, peanuts; in fact, most any crop grown by general farmers. Adjustable width, it is capable of cultivating any row from two to four feet. Pivotal device for wheels enables quick dodging in crooked rows and listed corn. Shovel equipment can be furnished to meet all requirements.

NA Two-Row Cultivator.

Will save money. It cultivates two rows at a time. One man can handle it as easily as a single-row cultivator. All steel, well trussed frame, there is no sag. Dodge is quick and easy. Three or four horses can be used. Remember too, corn rows are always in full view.

JOHN DEERE ALFALFA AND DISC CULTIVATORS.

Alfalfa Cultivator.

Has been constructed especially for alfalfa, breaking up the top layer of soil, which becomes packed and hardened. Lighter draft and cultivates more thoroughly in alfalfa than any other machine. Tooth-bars are hung independently in frame and controlled individually by spring pressure.

No. 23 Disc Cultivator.

Is neat in appearance and simple in construction. Steel and malleable have been used throughout. It is light and strong. Gangs are angled instantly by means of lever and rack. They are locked securely. It is unnecessary to take off gangs when changing from in-throw to out-throw, or vice versa.

No. 28 Disc Cultivator.

Frame has strength and stiffness. Does not spring, but holds rigidly. A heavy and high arch is used, which makes it stronger and capable of use in high corn. Axles can be adjusted without disturbing rest of machine. It has pivotal pole with foot treadle dodge.

No. 29 Disc Cultivator.

Pole and seat bow are made as one unit and pivoted to arch and main frame, parallel draw-bars keep disc gangs running level, doing uniform work and giving gangs independent rise and fall. Gangs are instantly adjusted without use of tools. Shovel and spring tooth attachments can be furnished, making this a universal cultivator.

Circa 1860 - 1895

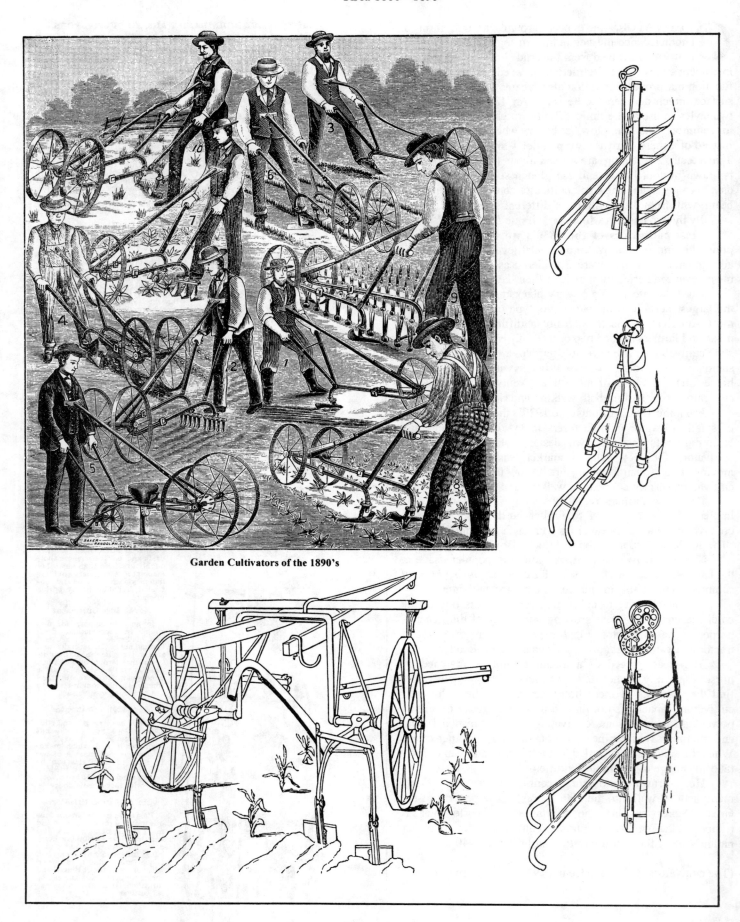

Garden Cultivators of the 1890's

B. F. Avery & Sons, 1915

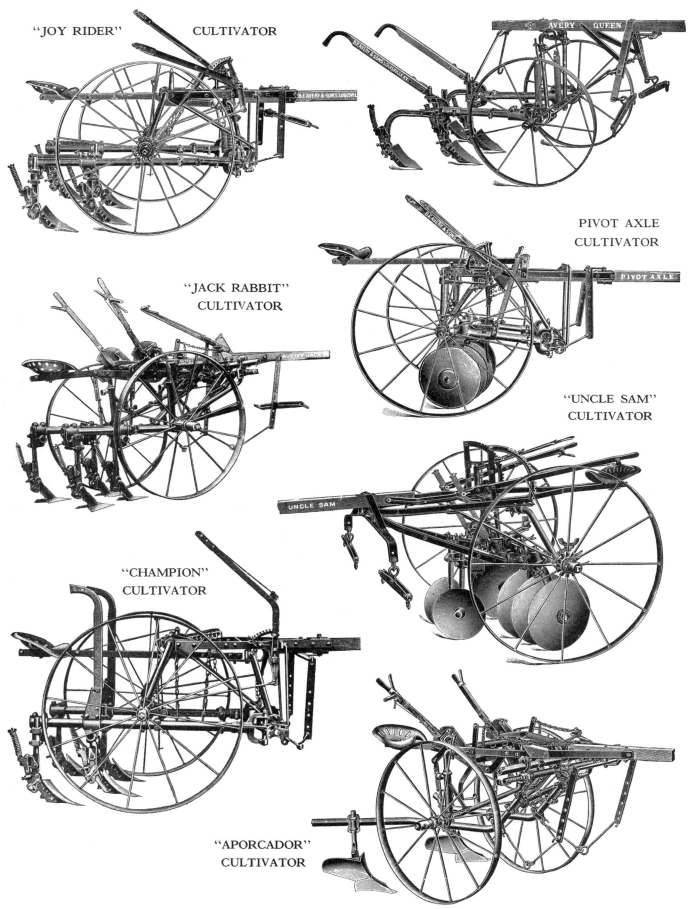

AMERICAN AGRICULTURIST

VOLUME XXXVII.—No. 5. NEW YORK, MAY, 1878. NEW SERIES—No. 376.

THE CHAMPION COMBINED MOWER AND SELF-RAKING REAPER.—*Drawn and Engraved for the American Agriculturist.*

Until recently, combined reapers and mowers have been considered unsatisfactory machines, and farmers have been to some extent prejudiced against them. This has been unfortunate, because few farmers can afford to buy two machines to perform two similar operations, when one could be made to serve the purpose. But after six years of successful work it has been proved that the mechanical difficulties in the way of the effective operation of a combined machine have been vanquished by the manufacturers of the machine of which we give the above illustration. This is the Champion Combined Self-raking Reaper and Mower, made at Springfield, Ohio. The character of the machine is readily apparent from the engraving, and the full details, which show the reaping-machinery added to the mower, are given on another page. At figure 1, on page 168, the mower is shown so clearly, that there can be no mistake about the parts, and at figure 2 the added apparatus is so shown, that the operation of the combined machine is equally clear. It is needless to say anything in regard to the reputation of the Champion machines. The name is familiar to every farmer, who has seen or heard of mowers and reapers, and the fact that 35,000 of these machines are made annually in the workshops of the Company—the largest manufactory of agricultural implements in the world—evidences the estimation in which they are held. In regard to awards for excellence at exhibitions, the Champion machines have stood in the front rank of those exhibited. It took five first premium medals at the Centennial Exhibition, and an award of five diplomas for the finest exhibit in the Hall and for merit in the great field trial held in connection with the Centennial. At this trial the Champion machine recorded the remarkably light and hitherto unexampled draft of only 131 pounds. It also succeeded in cutting perfectly grass that had been beaten and laid by storms, and in addition, to make the test more severe, had been flattened by a heavy roller drawn over it. No severer test than this could be imagined. As a reaper it cuts the grain successfully, although it may be laid and tangled in the worst manner, and delivers it in good and even gavels ready for binding. These two important operations are accomplished equally well, and the change from mower to reaper or back again can be made with ease in twenty minutes. Clover-seed and flax may also be successfully harvested with it. The catalogues of makers of first-class machinery are publications of high excellence, and serve a purpose beyond mere advertising. To this the catalogue describing the Champion is no exception, as the machine is shown in action in its various combinations and positions, including its appearance on the road, while every detail of machinery is given with such distinctness and accuracy, that one at all acquainted with mechanics can, by a study of the catalogue, understand the construction of the machine, and form an opinion of its working in the field, almost as well, as from an actual inspection of the thing itself.

HARVESTING TOOLS followed a path of gradual evolution, starting with the short-handled flint or bronze *sickle* of 1500 B.C. From this tool Flemish farmers developed the Hainault scythe. It had a broad two-foot long blade and a 12 inch crank-shaped handle with a leather finger loop at its end. Later the handle was lengthened to four or five feet and the Hainault grain cutter evolved into the traditional long-handled *European scythe*.

In the 16th century, three cradle-like curved wooden fingers were added above the scythe blade to catch the stalks, thus allowing the "cradler" to tilt his modified scythe and distribute the grain more evenly on the ground for another worker to rake and tie into bundles. The five-fingered *American cradle* appeared in the 1750's. Expert cradlers were in great demand and made twice the daily wage of ordinary farm workers. A first class cradler, trained from boyhood, could cut three or four acres a day, while an average farm laborer did well to complete an acre of work.

A Victorian lady recalled: "I remember the reapers who appeared one day – big, sweaty, wholesome men, carrying their cradles. They were soon at their labors in the field, sweeping steadily onward with powerful arcs of their long, sharp blades. The wheat sliced off in bundles. Somebody followed along behind, to tie it all up carefully, piling the sheaves in neat stacks, but leaving some for the gleaners."

Wheat could not be allowed to stand in the field more than ten days before it began to shatter from the heads, and before the invention of reaping machines huge quantities of grain were lost annually due to a chronic shortage of labor during this critical time period. In the great plains, as harvest time approached, farmers began to canvas nearby towns for helpers to reap, bind and stack their crops. Everyone from lumberjacks to store clerks and school children were recruited to bring in the crop before birds, rodents, insects, or a sudden storm could destroy it.

Mechanical reapers, mowers and **grain binders**, that were developed between 1830 and 1880, gradually alleviated the problem of seasonal labor shortages. The owners of these new implements often worked fourteen hours a day, traveling from farm to farm at peak periods of the harvest season.

Mechanical *reapers* were used to cut grain crops, and *mowers* to cut hay. Reapers either delivered the grain stalks to one side in small clumps called gavels, for a following worker to bind up into sheaves; or, in the case of the circa 1858 *harvesters*, they raised the stalks onto an attached platform where another operator bound them by hand.

Hand binding of wheat was replaced in 1875 by a patented machine with steel fingers that wrapped wire around each bundle and twisted it into a finished knot. However, many farmers objected to the use of wire because of its danger to feeding cattle. The hazardous wire binders were replaced by John F. Appleby's cheaper and safer twine-binder, manufactured by William Deering in 1879.

The McCormick and Deering firms were the two leading manufacturers of *grain binders* in the 1870 - 1890 period. One 50,000-acre farm in North Dakota used 282 self-binding reapers to bring in its 1880 harvest.

The first mechanical reapers date back more than 3,000 years. Pliny observed that some farmers in ancient Gaul fixed a series of knives to a cross bar on the tail of a two-wheeled cart and drove it through the grain, clipping off ears or heads as it went. Many efforts to achieve the same result without damaging the grain were tried over the next several hundred years. One such invention was a cart with comb-like blades mounted in front of the bed, which cut the grain as it moved through the field, ahead of a pair of oxen.

In 1577, the French writer, Barnaby Googe, mentioned a reaper consisting of "kind of a low carriage, equipped in front with sharpened sickle blades and cutting everything before it." In 1788, a French landowner presented to the Academy of Science a new machine pushed by a horse, with six revolving scythes on the front, that would cut 7,000 square feet of grain an hour.

Fifty-five reaper patents had been issued in France, England, Germany and the United States before 1831. However, most of the machines that were actually built did not function well enough to be marketed as practical harvesters. The following British inventions paved the way for the ultimate design of successful American reapers and mowers.

An Englishman, Thomas Gladstone, pioneered the idea of a side-cutting machine with a revolving cutter, and inside and outside dividers, in 1806. The grain fell upon a platform that was cleared with a hand rake.

In 1822 Henry Ogle, a schoolmaster from Remington, built a machine with a reciprocating knife, driven by cogs on its wheels, and the first reel used on a reaper.

BELL'S REAPING MACHINE

The most successful reaping machine developed prior to 1833, was invented by an English minister named Patrick Bell. His push-style reaper, introduced in 1826, had a bank of oscillating knives, 15 inches long and about 4 inches wide at the back, that worked over a similar set of knives underneath them like so many pairs of shears. The rear of each movable blade was attached to an oscillating rod that was connected to a worm flange on a revolving shaft. Bell's machine presented a new idea: a *canvas conveyor* that moved on rollers and carried the grain to one side where it was deposited in a continuous swath. Bell also provided his reaper with a reel and dividers. Even though it was not a practical implement, being pushed rather than pulled, Bell's reaper was used in England for several years until replaced by American machines invented by Hussey and McCormick.

Hussey's Reaping Machine of 1833

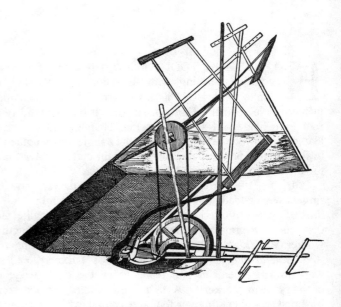
Mc Cormick's Reaping Machine of 1834

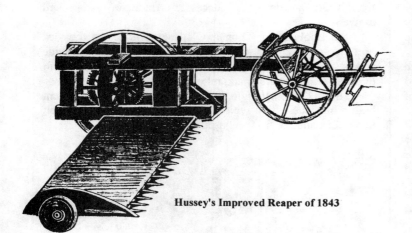
Hussey's Improved Reaper of 1843

Mc Cormick's Improved Reaper of 1847

Hussey's Improved Reaper of 1847

Ketchum's Mower of 1847

OBED HUSSEY, of Baltimore, Maryland, patented a mechanical reaper in 1833, which, with various changes, was used for the next fifty years in the western United States. Hussey described his machine in a July, 1843 issue of the *American Agriculturist:*

"When in operation the horses travel on the stubble near the standing grain, drawing the machine behind them. The part which cuts the grain is a wide platform extending six feet into the grain, and it is capable of any adjustment, from five to fifteen inches from the ground. Along the forward edge of the platform is a row of strong iron spikes, formed of two pieces, one above the other, with a horizontal slit in each spike for the cutting-blades to play in. A crank, turned by a cog-wheel connected to the main axle, gives a vibratory motion to blades, causing them to move out of one spike and into the other. As the grain is cut, it falls backward onto the platform."

"When the wheat is tangled, this falling back is aided by an instrument in the hands of a man riding the machine, whose business it is to push off the grain in heaps as it accumulates on the platform. He is able to do this with great accuracy and neatness, leaving the heaps in fine order for the binders. One machine will cut twenty acres a day with ease and the blades need no sharpening from beginning to end of harvest. Ample proof of these facts can be had from five hundred farmers in the United States. The price of my machine with shafts for a single horse is $150. The forward wheels are an extra $20."

Cyrus McCormick had also been at work refining his reaper since 1831, but he neglected to patent it until 1834, a full year after Hussey went public. McCormick was sure that his invention of the reaper took precedence over Hussey's patent and a battle between lawyers and reporters representing the two sides went on for several decades. Multiple suits were also filed by both parties against other manufacturers who had crowded into the lucrative reaper market.

Of the two machines, perhaps Hussey's was the most improved over earlier devices, and it was nearer in design to later successful models. At first the honors were evenly divided, however, Hussey did not have the energy and perseverance of McCormick, and eventually lost the struggle for market supremacy.

During the 1840's and 1850's, McCormick made several improvements on Hussey's design and aggressively pursued his goal of worldwide distribution. In 1847 and 1848 McCormick added a seat for the raker, a cutting apparatus, reel divider, and a platform. By 1849, another seat had been added for the driver, and in 1851 a sickle, made in sections, replaced the straight knife.

Palmer and Williams patented a sweep rake in 1851, that swept the platform at regular intervals, leaving the grain in bunches on the ground for binding. Production of a combined reaping and mowing machine with the patented automatic gavel delivery attachment followed from 1862 to 1873.

In the summer of 1855 American machines took first and second place in trials at the Paris exposition by cutting an acre of oats in half an hour. Both models could be could be converted from reapers into mowers in a few minutes.

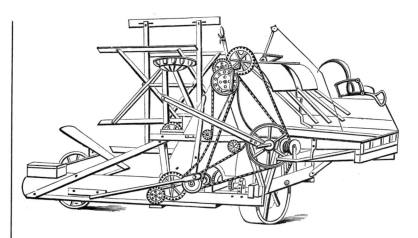

The McCormick harvester and wire binder (above), went into commercial production in 1877 and eliminated the hand-binding platform. Automatic twine-binding replaced wire binders in 1881.

The McCormick "Folding Daisy" style reaper was manufactured from 1882 to 1902. Its rakes not only held the standing grain against the cutter bar, but also raked the cut grain from the platform in any size desired. The platform and rakes could be folded for faster road travel and passing through narrow farm gates, (see illustration on next page).

Cyrus McCormick was born into a family of prosperous farmers in 1809. His father was an inventor of sorts, and owned a fully equipped blacksmith shop on his twelve-hundred-acre Virginia farm. For twenty-five years, off and on, the senior McCormick had tried to devise a practical mechanical reaper but he had not brought his invention to the point of actually reaping a crop. Finally, in 1831, after cutting a field of wheat with his machine, only to find that it left the stalks too tangled to bind, Robert McCormick threw in the towel and quit the inventing game.

The younger McCormick immediately took up the challenge, and a few months later, at the tender age of twenty-two, demonstrated his new four-horse machine by cutting six acres of oats adjacent to a local tavern. This first effort damaged some of the grain and three more years of refinement were necessary before McCormick secured a patent in 1834.

The essential difference between his reaper and twenty other machines patented before it, was that previous models employed fixed sickle-shaped cutters. McCormick's design resembled a reel-style lawn mower with four horizontal gathering blades that moved reciprocally with the forward motion of the machine. The ground-level cutting bar had an edge like a sickle and worked against a set of wires, which substituted for the iron cutting-spikes of Hussey's machine. A bar divided standing stalks from the row being cut and the grain fell upon a platform where it was raked to one side by a man walking along beside the machine.

The financial panic of 1837, brought on by excessive land speculation and a collapse of the cotton market, caused Cyrus to lose his farm and foundry. However, the depressed economy motivated him to aggressively pursue the final development of his reaping machine for the marketplace. By subcontracting the actual manufacture to others, McCormick

managed to produce and sell two reapers in 1841. Word of their efficiency spread and during the next thirty-six months he was able to sell eighty machines at over one hundred dollars each.

In 1847, at the age of thirty-six, Cyrus moved to Chicago where he talked the mayor of the city into putting up $50,000 for one half interest in a factory venture. The area around Chicago was extremely fertile in more ways than one. McCormick sold a thousand reapers in 1851 and bought out his partner. By 1857 he was moving 23,000 machines a year out of the factory doors, and during the season of 1860 McCormick sold 4,000 reapers to Chicago area farmers alone. An 1850 account of McCormick's reaper in use on an Ohio wheat farm follows:

"They had hitched four horses to the machine, although two horses are ample; the leaders keep the tongue straight on the line of draft and tend to steady the machine. Two men ride on the implement, one to drive, the other to rake off the grain and leave it in heaps for the binders. One machine cuts from 16 to 18 acres a day and gets the crop in on time, which could not be done with hand labor, which is presently impossible to find, even at $1.50 to $2.00 a day."

During the great Chicago fire of 1871, the McCormick Reaper Works burned to the ground. Cyrus was wealthy enough to retire but his young wife implored him to rebuild, saying: "I don't wish for our sons to grow up in idleness." Thirteen years later his oldest son, Cyrus Jr., became president of the company at the age of twenty-five, and in 1902 he supervised the merger of the McCormick Reaper Works with Charles Deering's factory, and four smaller competitors, that became known as the International Harvester Company.

The **Deering Harvester Company** had evolved from a merchandising partnership in 1870 between William Deering (Charles' father), and Mr. E. H. Gammon, a Chicago area businessman. Deering and Gammon were the distributors of C.W. Marsh's grain harvesters. In 1879 Deering bought out Gammon and joined forces with J. Appleby to begin mass-production of Appleby's twine binder.

Deering and McCormick had been bitter rivals for more than a quarter century. And the two firms controlled ninety percent of all grain binder and mower production in the United States in 1902, when they finally agreed on an equitable sharing of $60,000,000 worth of stock in the newly formed IHC corporation.

The new **International Harvester Company** soon purchased other implement factories in order to gain quick entry into the booming farm machinery market. The D. M. Osborne Co. of Auburn, New York, brought a proven line of haying machines into the fold and the acquisition of the Keystone Company of Rock Falls, Illinois, provided a popular brand of corn shellers, tillage, and haying tools. Later came the purchase of the Weber Wagon Works, the Kemp manure spreader, and Parlin & Orendorff's plows, planters, beet machines and potato diggers. In 1919 the Chattanooga Company's plows, cane mills, evaporators and syrup equipment became a part of the line and in 1920 American Seeding Machine's grain drill operation was added to the fold. Meanwhile, in 1907, an empty factory complex in Akron, Ohio, provided IHC with needed manufacturing space for the companies' new high-wheeled farm trucks, the forerunner of today's line of International trucks. Tractor production began before World War I.

The Marsh Harvester of 1858 was the first step toward the first self-binder manufactured seventeen years later. In 1877 this was superseded by the "wire" binder, which in turn was displaced by the Deering twine binder in 1879.

Marsh Harvester

THE MARSH BROTHERS, of De Kalb, Illinois, patented the Marsh Harvester on August 17, 1858. Previous to this time attempts had been made to build harvesting machines that would bind the grain before delivery to the ground, but none were commercially viable. In 1860, when the Marsh harvester began to seek a niche in the market, reapers, hand-rakers, self-rakers, and droppers were the only harvesting-type implements available.

The Marsh brothers did not start entirely from scratch. On June 19, 1849 a patent was granted to the Manns for an attachment to McCormick's reaper that carried the grain from the ground-level cutter up an endless belt to a bundle-packing mechanism before dropping it to the ground for binders to pick up and tie.

The theory behind the Marsh brothers' modification was that two men, standing on the harvester, could bundle and tie the grain while on the move; but only if it was conveyed to them at a practical working level in good condition for binding.

The new machine was designed so that the grain would be delivered at an elevated height to a table on the platform, where two riding attendants, standing back to back, would receive the grain, bind it into sheaves, and toss it on the ground in an orderly fashion.

The two brothers, along with J. T. Hollister, organized a company in 1864 and built 24 machines during their first year in business. By 1870 the plant was turning out a thousand machines annually, and eventually was acquired by the Deering Harvester Company.

Other contributors to the further development of harvesting machines in the 1870's were: George Spaulding, who invented the packer; Sylvanus Locke who designed the steel fingers that wrapped wire around the bundle and knotted it; John Appleby, who refined the packer and invented the first successful twine binder; and Jonathan Haines, who invented a machine for heading grain and elevating it into a wagon (see "header" illus. on page 63).

All *self-binding harvesting machines* that followed between 1881 and 1885 were based on the Marsh harvester's design. A farm machinery textbook of the period enumerated its features. "The modern *self-binding harvester* consists of: (1) A drive wheel in contact with the ground. (2) Gearing to distribute power from the driver to the various parts. (3) The cutting mechanism; consisting of a serrated reciprocating knife, driven by a pitman from a crank, and guards or fingers to hold the grain while being cut. (4) A reel to gather the grain and cause it to fall in form on the platform. (5) An elevating system of endless webs or canvases that pass over an iron platform and carry loose grain to the binder. (6) A binder to form and tie the grain into bundles.

In the 1920's some farmers began converting from horse-drawn machines to tractor-driven *grain binders*. Speed was necessary to save all of the crop during the short harvest period. The cutting and binding mechanisms of tractor binders were driven directly from the tractor by a power shaft, thus eliminating the earlier delays caused by wet or loose ground when horse-drawn (ground-driven) binders were employed.

ADRIANCE, PLATT & CO. was founded in 1878, in Poughkeepsie, New York, where they engaged in the manufacture of high quality mowers, reapers, grain binders, hay rakes and harrows for the next thirty-four years. The company was subsequently purchased in 1912, by the Moline Plow Co., of Moline, Illinois.

The apron elevator binder illustrated above was first made in the 1890's and evolved into the 1907 model pictured Prior to 1902 the gathering reel was driven by chains and sprockets; but a shaft and gears were substituted as shown in the cut above. In 1903 major changes were made in the knotter; the twine disk being driven by a cam pinion instead of a ratchet.

The Adriance binder shown above was a less expensive, low-profile machine, built from 1888 through 1907. It was regularly made with a five- foot cutter, but was also available in a six-foot model. Its gathering reel also was converted from chain and sprocket to gear drive in the early 1900's. Adriance also made a low-profile rear discharge binder.

WALTER A. WOOD, of Hoosick Falls, New York, entered the reaper and mower market in 1853, completing about one machine a day during his first year in business. Ten years later Wood's factory was turning out 6,500 units a year. Between 1855 and 1875 more than 230,000 Walter A. Wood reapers, mowers, and harvesters were sold. The Civil War had produced a labor shortage in many areas of the country, and Northern farmers were buying mowers and reapers in record numbers. Meanwhile, in the South, Sherman's army had destroyed every piece of agricultural equipment they found, and many southern farmers had lost everything but the clothes on their backs. The eventual recovery in the South was a boon to many farm equipment manufacturers.

In reviewing trade publications of the era, we found more references to Walter A. Wood machines than any other brand. A full page advertisement in the February, 1877 issue of *Country Gentleman's Magazine* stated that Wood also had showrooms in London, at 36 Worship Street. The ad went on to declare that: "Walter A. Wood's mowing and reaping machines have, during the seasons of 1873, 74, and 75, competed in 234 field trials. In these competitions we have met and defeated forty different makers." Madison Avenue would be hard put to beat this kind of "down-to-earth" advertising today. (The engravings on this page are reproduced from an 1876 business directory.)

IRON FRAME MOWER FOR TWO HORSES.

HARVESTER AND SELF BINDER COMBINED.

WALTER A. WOOD'S HARVESTING MACHINES,
COMPRISING
Iron Mowers, Self-Rake Reapers,
Self-Rake Reapers and Mowers
Combined, Harvesters and Self-
Binders Combined.

IRON FRAME MOWER WITH MANUAL DELIVERY REAPING ATTACHMENT.

SWEEP RAKE SELF DELIVERY REAPER ON THE ROAD.

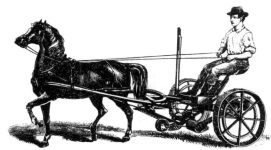
IRON FRAME MOWER FOR ONE HORSE.

HEADER.

THE ORIGINAL and only reliable Hay-Pitching Machine of its kind in the world. Is more simple in construction, easier of operation, better made, and more finely finished than any other Hay-Stacker. Elevates the hay higher, is of lighter draft, will do twice the work, and last three times longer.

"MONARCH" HAY RAKE

THE MOST COMPLETE SULKY SWEEP RAKE ON EARTH.

HAY HARVESTING TOOLS. Hay has to be cut, cured and removed from the field at just the right time or much of its nutritive value is lost. This was especially true of alfalfa, the oldest hay plant grown during the period of our study. Alfalfa came from Media to Greece in 490 BC. It had been grown in New York since 1800, but its most effective introduction into our agricultural economy was a variety brought from Chili to California in the 1850's. It was also well adapted to the warm climate of the Arizona Territory, where up to eight crops a year could be harvested if sufficient moisture was available.

Timothy was the most important hay plant grown in America. Account books kept by northeastern farmers show that in the years prior to 1915, timothy hay was the most profitable crop they could raise. The seed was cheap and a whole clump of grass came up from each one planted. Only ten or fifteen pounds of seed were required to sow an entire acre. Other popular varieties of hay were blue grass, red-top, red clover, oats, and orchard grass. Eventually the big city demand for horse feed dried up and by 1916 the market for timothy hay had gone into a downward spiral as trucks and automobiles began to replace horses on American roadways.

"*Horse rakes* are very useful on all smooth meadows. It is said, a man, a horse, and a boy, will gather hay with this implement as fast as six men with ordinary hay rakes. The expense of constructing a horse rake will not exceed two dollars. It is composed of a 3 x 3-inch piece of scantling 10 feet long, with 25 teeth, 1 1/2 inches in diameter, by 2 feet long. The teeth should turn up a little at the end, to prevent their digging into the earth. Eight pins, 24 inches in length, are driven perpendicular into the scantling, and another light piece at the top. Also attached are two plough handles. In operation, the teeth run along the ground under the hay, and as they take it up, the upright slats retain it till the rake is full; then a man turns it over, and empties it in a row."
– From an article in *The Farmer's Guide,* of 1824.

Sweep rakes and *push rakes* evolved from these early wooden horse rakes and were used well into the 20th century, especially in areas where field stacking of large amounts of hay had to be accomplished in a short period of time. The sweep rake had straight wooden teeth to take hay from either a swath or windrow, and was either drawn between two horses or pushed ahead. (An 1888 model is shown above.)

When a load was secured the teeth were raised, the load hauled to a *stacker* and placed upon its teeth, and the rake backed away. Early *hay stackers* were made in two general styles: the *overshot* and the *swinging* type stacker.

In the overshot the teeth were drawn up and over, thus throwing the load directly back upon the stack, with the help of horses, ropes and pulleys.

The *swinging stacker* permitted the load to be locked in place after being raised from the ground, and swung to one side over the stack, and dumped. Then the fork was swung back and lowered into place to receive another load. These tools were all of solid wood construction.

Metal tined *horse rakes* began to appear in the late 1840's and the *sulky rake*, or *dump rake*, came into use soon after mowers were refined in the 1850's.

Walter A. Wood was the first to manufacture a spring-tooth rake, circa 1853. These early dump rakes had wooden frames and wheels, and a hand-powered lift lever that raised the steel tines to deposit hay in a windrow.

Later rakes were made entirely of steel and the dump mechanism became a power-assisted operation — first by a trip lever, then further modified by a foot clutch. The introduction of hay loaders created the need for a longer, lighter windrow, and *side delivery rakes* became popular in the 1890's; however, dump rakes remained in common use for several more decades.

When hay is mowed its water supply is cut off, but a lot of moisture remains in the plant that must be reduced to a safe percentage for storing. It was thought that if the leaves were prevented from drying up too rapidly they would aid in carrying off moisture from the stems. Green clover contains about 85 percent water. When properly cured about 25 percent of this water remains. After falling to the ground the leaves, or tops, are exposed to direct sunlight and if left too long the leaves will shatter easily. High-grade alfalfa hay must retain its green color, be fresh and sweet, and keep most of its leaves on the stems during handling

In the 1840's, horse-drawn two-wheeled *hay spreaders* with revolving tines, were introduced to stir up the hay from the swaths and allow it to cure more evenly. In the 1870's these implements were renamed "*tedders*" and became widely used until side delivery rakes became popular in the 1890's. However, some farmers kept on using *tedders* till the 1950's.

An 1870 mower, working with the cutter bar elevated, and the right wheel passing through a ditch.

HAY MOWERS appeared on the market in the 1830's, but sickles, scythes and cradles continued to be used on many small farms until the end of the 19th century.

Obed Hussey's first reaper was really a *mower*, and its design formed the basis for future mowing machines. As previously mentioned, many of the early machines were combination reapers and mowers that could be adjusted to cut either grain or grass.

William Ketchum has been called the father of the mower trade because he was the first to market mowers as distinctly different machines than reapers. He took out several mower patents, including one in 1847 that featured an unobstructed space between the driving wheel and the side-mounted cutting bar. (See illustration on page 58.)

The first mower with a flexible cutting bar was invented by Hazard Knowle; a Patent Office machinist who is also noted as the inventor of an iron woodworking plane in 1827.

Cyrenus Wheeler was to the mower, what Cyrus McCormick was to the reaper. His patent of 1854 marked the division between the two machines and combined so many important features that it gave him a lot of credit for the development of the mower. But it was not until 1858, when Lewis Miller added several new features, that a really practical two-wheeled metal mower called the "Buckeye", came on the market. A. Kirby and E. Ball made other valuable improvements and by 1860 the "modern" two-horse mower with its four to eight-foot cutting bar was perfected.

Great American Industries, a history book published in 1872, mentions the Clipper mower: "In the great trial held at Auburn, New York, in 1866, fifty-four harvesting machines were on exhibition, and few produced a more ingenious mower than the 'Clipper', an invention of Mr. Rufus Dutton. It is made of the finest materials and combines safety for the driver with the convenience of management and adaptation to uneven surfaces. Over 12,000 of these machines have been sold and hardly one has required replacement from breakage."

"In the four or five years past, from eighty to ninety thousand machines (of all brands) have been made annually in the United States. Among the most prominent are those establishments making the 'McCormick', the 'Buckeye', the 'Kirby', the 'Wood', and the 'Clipper' mowing machines."

Space only permits us to mention of a few of the dozens of mower patents issued between 1835 and 1865. Victorian lawyers must have grown rich filing and defending law suits initiated by various manufacturers during this period of overlapping patent claims.

The next step in the evolution of mowing machines was the displacement of horses, and the ground contact drive wheel, as power sources. Greater mowing capacity was added by the tractor-powered cutter bar and a slip clutch that released automatically when the sickle clogged, or the strain of cutting tough material was to great.

When the cutting bar of a 1930's tractor mower hit an obstruction, the bar swung back, the power shaft telescoped, and the clutch automatically disengaged and stopped the sickle. During ordinary field operation, a foot-lift raised the cutter bar high enough to avoid most rocks and tree stumps.

Side-Delivery Hay Rake THE COLUMBIAN.

The only successful rake delivering in continuous windrow in line with the work of the mower. . . . .

Perfectly meets a growing need of the times. . .

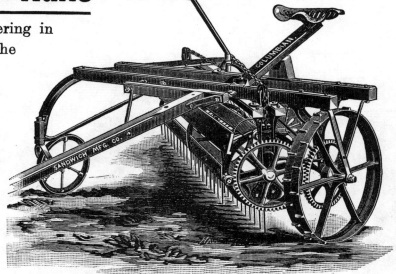

SPRINGFIELD HAY RAKE
The only successful SELF DUMP RAKE.

Any boy can operate it.

Simple in principle and perfect in construction. It has been on the market for three years, and has been thoroughly tested. Every rake warranted. Illustrated Circulars free. Address the
SPRINGFIELD MAN'F'G CO., Springfield, Ohio.
Mention the Farm, Field and Fireside.

COATES' "LOCK LEVER" HAY and GRAIN RAKE.

Patented Aug., 1877; Jan., 1875; June, 1875, and Nov., 1876.

97,000 NOW IN USE.

Twenty Steel Teeth. No ratchet wheels, friction bands, nor other complicated machinery needed to operate it. Slight touch of lever and DRIVER'S WEIGHT dumps it. Best self dump in market. The best and easiest working rake in the world. A small boy rakes easily 20 acres per day with the COATES' "LOCK LEVER." Send for Circulars.

A. W. COATES & CO.

THE RACINE WIND MILL

THE LIGHTEST RUNNING AND MOST PERFECTLY SELF-REGULATING MILL MADE. PUMPS AND TANKS OF EVERY DESCRIPTION.

Agents Wanted. Send for Catalogue.

WINSHIP MF'G CO., RACINE, WIS.

☞ MENTION FARM IMPLEMENT NEWS.

REID'S PEERLESS CREAMERY
Absolute Perfection for Best Quality Butter.

BUTTER WORKER
MOST EFFECTIVE and CONVENIENT.
Also CHURNS, POWER BUTTER WORKERS, PRINTERS, SHIPPING BOXES. Send for my Illus. Catalogue, containing valuable information for Creamery men and Butter Factories.
CREAMERY SUPPLIES.
A. H. REID, 30th and Market Streets, PHILADELPHIA, PA.

The Grand Rapids Hay Tedder

Manufactured by the
GRAND RAPIDS MFG. & IMPLEMENT CO.,
GRAND RAPIDS, MICH.
TO MAKE HAY SUCCESSFULLY YOU MUST USE A TEDDER.

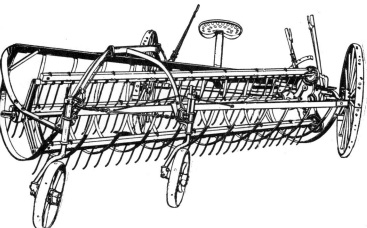

SIDE DELIVERY RAKES similar to the engraving on the opposite page were perfected in the 1890's. By the 1920's they had evolved into the traditional style pictured above. Their function was to lift the hay and place it in fluffy windrows with the green leaves inside, protected from direct sunshine. The leaves, thus shaded by their stems, cured rapidly to a uniform green color by the free circulation of air through the windrows. The hay was turned over once or twice during this curing process to keep it fluffy.

There were three types of loaders, the *single cylinder*, the *double cylinder*, and the *raker-bar*. The single cylinder, or *endless apron* style was used only for loading directly from a single windrow. Its cylinder reel had eight bars with rake-like teeth that projected downward, when they were nearest the ground to pick up the hay. As the loader moved forward the cylinder revolved and the teeth turned upward, delivering the hay to the carrier where it was elevated to the top of the wagon.

The double cylinder loader — with its floating gathering cylinder — could be adapted to load from a swath, or a single, or double windrow. When properly adjusted, its counter-revolving reels could do better job of gathering hay in rough fields than the other two types.

The earlier raker-bar, or "fork loaders", were easier to operate, having a simple lever to adjust the height of the ground rakes, which placed the hay upon an elevating apron. They could be used to take hay from either a swath or a windrow. Later models of raker-bar loaders had drive chains that moved a series of bars with rakes on their ends in oblong strokes which lifted and pitched the hay to the forward end of the wagon, making it possible for one-man loading.

HAY LOADERS were developed by The Keystone Mfg. Co. as early as 1875 and reached the marketplace in the 1880's. The first hay loaders were designed to be attached to the rear of a wagon, to gather hay from the ground and elevate it to a rack on the wagon. They were one of the best labor saving machines a farmer could buy. Grateful owners remembered the "not-so-good-old-days" when they spent all day tossing loose hay from the ground to the top of a wagon with pitchforks...a slow, hot, dusty, sweaty job. The hay loader was needed on every farm where hay was gathered in the field and transported to a hayloft, or hauled to stacks for storage. It took only ten or fifteen acres of hay to justify the purchase of a hay loader — which was an ideal companion tool to the side delivery rake — but most farmers put off the purchase of mechanical hay loaders until the early 1900's.

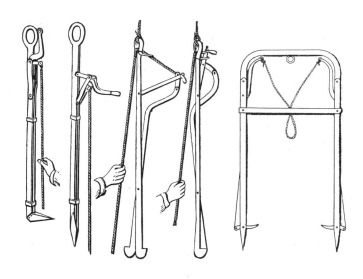

HAY FORKS. The introduction of new hay-harvesting and stacking tools created a market for other tools to move large quantities of hay in and out of the barn. The *harpoon horse hay fork* was patented by Edward Walker in 1864.

D. B. Rogers and Sons bought Walker's patent and made two or three improvements, than began to manufacture harpoon forks on a grand scale. Unfortunately, Rogers did not due enough field-testing, and all of the forks he had sold were returned by irate farmers who demanded refunds because the tripping mechanism failed to release properly.

Rogers dropped the hay fork like a hot potato, and another inventor, A. J. Nellis, bought all the rights and added twelve more patented improvements over the next few years.

By 1875 Nellis and Co. had won seventy-six state fair prizes and sold 100,000 of the redesigned hay forks, which by then bore his name. (See Nellis double harpon fork above.)

HAY PRESSES attracted the attention of a number of inventors during the early 1800's and several patents were applied for.

In 1853, H. L. Emery, of Albany, New York, began to manufacture a machine with a capacity of five 250-pound bales an hour. It required two men and a horse to operate and produced a huge bale measuring two feet by two feet, by four feet. The next practical hay press to arrive on the market was invented in the 1860s, by P. K. Dederick.

In 1866 George Ertel developed a vertical *hay press* which was manufactured and sold in the West for several years. (Ertel's 1887 models are illustrated above and at right.)

The technology progressed, but it was not until the late 1880's that machines were available which could compress hay into a bale that could be tied and stacked successfully. These early presses required a two or three man crew and a couple of horses, or a steam engine, for power.

The term *hay baler* applies to the portable field machines, introduced in the late 1940's, that completed the whole baling task on the run, with just one operator.

The small, box-type, plunger presses were powered by hand, or by horses walking in a circle and pulling a sweep-arm drive. Larger hay presses, such as the 1910 "Wolverine" and the "Steel King" (above), were mounted on a wagon-style chassis that weighed up to 5,000 pounds. They required a five horse power engine and could kick out a bale a minute if the tying crew was fast enough to keep up. Other early 1900's plunger-style presses were being built with a capacity of 90 tons a day.

The obvious purpose of compacting and baling hay is to make it easy to handle, ship and store. There are about five pounds of loose hay in a cubic foot, prior to compression and the first machines were capable of pressing this loose hay into bales weighing 15 to 30 pounds per cubic foot.

By the late 1920s the weight of a cubic foot of baled hay had increased to forty pounds and the average sized bale weighed 75 to 100 pounds.

In the late 1930s and early 1940s, tractor-powered *hay balers*, that could gather, bale and tie hay, while on the run, became affordable on the average farm.

HAND-DUMP and SELF-DUMP PATTERNS.

OVER 100,000 IN USE.

ITHACA PORTABLE ENGINE
Economical, Strong and Safe.

ITHACA Broadcast SOWER
Complete in itself, or as Attachment to Rake.

SUPERIOR GOODS AT LOW PRICES

AGENTS WANTED in unoccupied territory. Address the manufacturers,

WILLIAMS BROTHERS,

Mention this Paper.   ITHACA, NEW YORK.

Satisfaction Guaranteed.

Most Simple and Rapid, Most Durable and Strongest.

ALL STEEL, FULL CIRCLE, CONTINUOUS TRAVEL

Hay Press.

Kansas City Hay Press Co.,
Kansas City, Mo.

WRITE FOR CATALOGUE AND PRICES.

The Ertel VICTOR
AUTOMATIC FOLDER,

Double-Acting Perpetual Hay and Straw Press.

A Machine Imitated, but not Equaled, Baling Hay or Straw faster, more compact, easier, more economically (to load 10 to 15 tons to the car), than is done with any other. So warranted or no sale. Circulars mailed free.   Address GEO. ERTEL & CO., Quincy, Ill.

Pumping and Power Wind Mills adopted by R. R. Cos. for superior workmanship and governing qualities. Agents wanted. Send 10 cts. for mailing catalogue and terms to agents.
American Well Works
Aurora, Ill., U.S.A.
Mention this Paper.

RICHMOND CITY MILL WORKS
RICHMOND, IND.
Manufacturers of
MILL STONES
FLOURING MILL MACHINERY

SPRINGFIELD, OHIO.
MANUFACTURERS OF THE

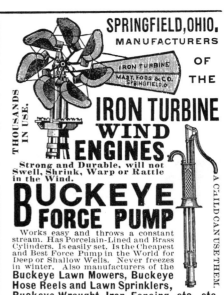

THOUSANDS IN USE.

IRON TURBINE WIND ENGINES

Strong and Durable, will not Swell, Shrink, Warp or Rattle in the Wind.

BUCKEYE FORCE PUMP

A CHILD CAN USE THEM.

Works easy and throws a constant stream. Has Porcelain-Lined and Brass Cylinders. Is easily set. Is the Cheapest and Best Force Pump in the World for Deep or Shallow Wells. Never freezes in winter. Also manufacturers of the **Buckeye Lawn Mowers, Buckeye Hose Reels and Lawn Sprinklers, Buckeye Wrought Iron Fencing,** etc., etc.
☞ Send for Circulars and Prices.
☞ MENTION FARM IMPLEMENT NEWS.

W. D. NICHOLS,
MANUFACTURER OF

Nichols' Celebrated Wind Mills,

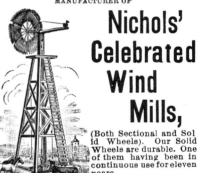

(Both Sectional and Solid Wheels). Our Solid Wheels are durable. One of them having been in continuous use for eleven years.

I also Manufacture

Feed Mills, Corn Shellers, Pumps, etc.

AGENTS WANTED. Address

W. D. NICHOLS, - - - Elgin, Ill.

Patch's Patent Corn Shellers.

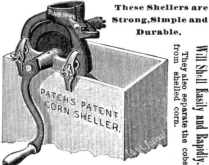

These Shellers are Strong, Simple and Durable.

Will Shell Easily and Rapidly. They also separate the cobs from shelled corn.

Twelve packed in a barrel for shipment.
Factory Retail Price, $3.00.

A. H. PATCH, Clarksville, Tenn

Bristol & Gale Co., Agts., Chicago, Ill.
David Bradley & Co., Agts., Minneapolis Minn.

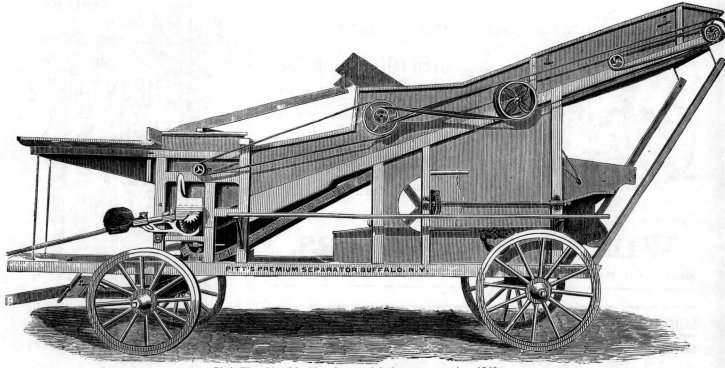

Pitt's Threshing Machine, four to eight horse power, circa 1860.

THRESHING MACHINES. Just a few years after the 1833 invention of the reaper, wheat production began to increase by leaps and bounds. In 1840 American farmers harvested 85 million bushels of the golden grain. By 1850 the figure had doubled, and in 1890 it doubled again to become one sixth of the entire world's wheat output.

These mountains of grain were separated from their stalks by a process called *threshing*, which for the previous 3,000 years had been accomplished by hand-beating with wooden flails, or by animals treading directly on the stalks or by pulling stone rollers over them.

Flailing grain on the barn floor was the commonest method of threshing until as late as 1850. A *flail* is simply a wooden club attached to a long hand-staff with leather thongs. The long handle allowed the farmer to flail away at the grain without bending over too far. One man could thresh 10 or 15 bushels of wheat a day while using a flail as illustrated above.

A farmer of the 1860's wrote: "I have threshed a great deal of grain in my day, and the average quantity of various kinds of grain that an ordinary laborer can thresh and clean in a day are 7 bushels of wheat, 8 bushels of rye, 15 bushels of barley, 18 bushels of oats, or 20 bushels of buckwheat."

After hand threshing, a cleaning process was necessary to separate the husks from the grain. This was accomplished by carefully raking the straw away, shoveling the grain and chaff into a basket, or a winnowing tray, and throwing the mixture up into the air to blow away the chaff.

Winnowing machines, called *fanning mills*, were first introduced in the 1780's. They consisted of a boxed enclosure containing a series of wooden paddle-shaped fans and several screens. A crank turned the fans and blew away the chaff as the grain fell through the screens.

The hand-threshing problem was addressed by several inventors in the 1700's; two Frenchmen, Duquet in 1722, and Rozier in 1766, and two Scottsman, Michael Menzies in 1750 (with a crude water-powered flail), and Andrew Meikle, in 1786. An Englishman, Mr. Lechie, from Stirlingshire, invented a machine with enclosed arms attached to a shaft in 1758; and a Mr. Atkinson, of Yorkshire, later devised a machine featuring a peg-toothed drum that ran across a matching set of concave teeth. This concept was used on almost every machine that followed.

The first practical threshing machine to appear in America was patented in 1837 by the brothers Hiram and John Pitts, of Winthrop, Maine. Their first crude machine consisted of a V-shaped wooden housing with a threshing cylinder mounted above an "endless apron" conveyor that carried the straw away to the top of a pile and allowed the chaff and grain to fall directly to the ground. However, it did not include a blower to winnow the grain, and sixteen more years passed before the brothers added a built-in separator. (The Pitts thresher illustrated above is circa 1860.)

In 1847 Hiram moved to Chicago, where, in 1852 he set up a separate Company, known as Chicago Pitts. John's factory, the Buffalo Pitts Company, was established in 1851, in Buffalo, New York, and lasted well into the early 1900's.

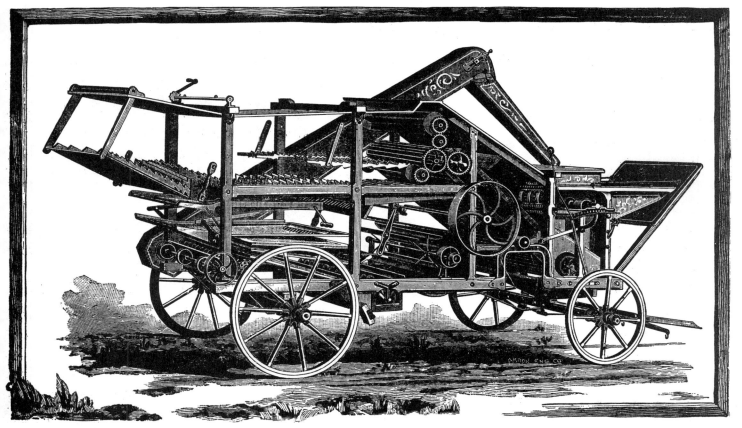

Interior view, Aultman "Star" Thresher, ca 1889.

THERE WERE dozens of competitors along the way, including George Westinghouse, Jacob Wemple, Cyrus Roberts, Nichols & Shepard, Aultman & Taylor, Meinrad Rumely and Jerome Case. By the 1880's steam had pretty much replaced horse-powered threshing machines and most of the above companies began to manufacture their own line of steam engines.

Threshing machines ended the same bottleneck in grain production that reapers had accomplished earlier. In the Paris exposition of 1855, several American threshing machines were pitted against six Frenchmen equipped with old-fashioned flails. In one hour the French laborers managed to beat out sixty liters of wheat. In the same time Pitts' American Thresher produced 740 liters vs. Clayton's English thresher at 410 liters. Duvoir's French machine only managed a feeble 250 liters an hour, which was still four times the output of six men.

The first threshing machines, which were initially called "thrashing machines", came in two separate units: The first unit, a *power supply*, came in two styles: a horse-drawn revolving shaft of the "sweep-arm" type, and the "railroad", a horse-driven treadmill with a power-take-off pulley.

The second unit was the *thresher* which consisted of a horizontal threshing cylinder, and (later) a blower fan. The two units were connected by a long leather belt that was lengthened by 25 feet in the 1870's because of the fire hazard posed by the newly introduced steam engines.

The first machines were made so that they could be loaded on trucks (sets of wagon wheels) and transported to a threshing site where they were removed from the trucks, and staked to the ground. These bulky machines were called "groundhog threshers". Later outfits, (like the 1889 Aultman "Star" thresher pictured above), were permanently mounted on wagon frames that could be easily moved from farm to farm during the busy harvest period.

In the 1850s, Pitts' thresher and separator was the machine most often used. It was operated by six horses hitched to a merry-go-round-style "sweep power" and could process 300 bushels of wheat a day. "This is done by men who make it a business, and go from farm to farm, having the help of two men at each place, and charge the owners three cents a bushel."

A typical threshing machine of the early 1900's required a 15 or 16-horsepower engine and could thresh from 500 to 1,000 bushels of wheat a day, or twice that number of oats.

It performed four distinct operations: (1) Shelling the grain from the head, by way of a toothed cylinder with matching concaves. (2) Separating the straw from the grain and chaff using a grate, a beater, a checkerboard, and a straw rack. (3) Separating the grain from the chaff and dirt, performed by the shoe, fan, windboard, several screens and a tailings elevator. (4) Delivering the grain to one location and the straw to another, which was accomplished by a grain elevator and a straw carrier, or stacker.

CASE THRESHING MACHINERY

NEVER in the history of the world has agriculture occupied a more preeminent position among industries than it does today.

While the world war is claiming the attention of nearly all nations, the whole world is looking to the farmer. It has been predicted by military strategists that the single item of food may be the determining factor in winning the war. This has a tremendous significance to the American farmer.

Today all countries are looking to the United States and Canada to feed the world. During the last year thousands of farmers answered the call of the Government to place more land under cultivation. The seriousness of the situation is best realized by the fact that the President of the United States found it necessary to appoint a food dictator to conserve the foodstuffs of this country.

What does this all mean? It means the greatest agricultural opportunity of centuries lies before the farmer. New lands must be opened. There must be more intensive cultivation. "Every useful implement should be bought and put to work" says a leading agricultural authority. For every bit of food is needed and more especially in these times of war.

In furnishing power farming machinery Case will play a most important part. In the great Civil War we were aiding in the food supply by furnishing threshing machines. That was nearly sixty years ago. We are still engaged in this great work, one that effects the life of all the people. While some are earning excessive war profits, we are going along as usual, building farm machinery that is productive of wealth and of the strength that protects when necessary.

Now England, France, Italy, Russia, Greece in search of the best have chosen Case machinery. The past year has brought to us thousands of new Case customers. These customers have come to us because other Case users have been good enough to tell their experiences with Case products. They told others why they would have none but Case. Our users are our best and most able salesmen, for they have put Case machinery to the test and they know of what they speak. Such enduring favor with practical farmers conclusively substantiates our belief that it pays to buy Case machinery. We are very grateful to Case customers who are working with us and for us. It is these things that spur us on to continue making Case machinery the very best on the market.

Jerome Increase Case was a farmer's son who had established a threshing business serving wheat growers in the upper Midwest. He built his first threshing machine in 1842, and founded a successful threshing machine factory in Racine, Wisconsin, in 1847. Within ten years Case was manufacturing 1,600 machines a year and by the turn of the century the Case Threshing Machine Company had become the world's largest manufacturer of steam engines.

Above, and on several following pages, we have reproduced illustrations from the J. I. Case catalog of 1918. At this point in time Case threshers were being built in seven sizes, from 20 x 28 up to 40 x 62, and were equipped with wind stackers and grain handlers. Steam threshing continued into the 1930's, when gasoline tractors became more economical. Improvements made to *combines* in the 1940's made all earlier forms of threshing machines obsolete by the 1950's.

Field hands on their lunch break pose with sacks of newly threshed wheat.

Kansas, 1912. Photo of water wagon, traction steam engine, and a Rumely thresher.

THIS STORY OF THRESHING in the "good old days" appeared twenty-five years ago in the Asheville, North Carolina, Times: "I remember how the threshing machine rolled down the road to our farm, hulking and cumbersome, pulled along by straining mules. The threshers pulled into the wheat field and set up there for business. I remember the tall stack of the boiler, with thick black smoke pouring out, and the towering frame of the threshing machine with its long wide belt, crossed in the middle. Dust and chaff were everywhere, it was hot dirty work. The machinery made a roaring racket and the flywheel emitted a steady whining sound. The men worked fast, tossing bundles in the hopper and stacking aside the sacks of precious wheat."

"**All work ceased when the dinner bell rang** and the men came in from the adjacent field. I recall the long harvest tables, covered with checkered red and white cloths, set for the men under the trees in our yard by the back door stoop. The big tables were loaded down with freshly prepared food; great stacks of biscuits, rounds of butter, jars of honey and home-made jam. There were sugar-cured hams, crisp fried chicken and steaks drowned in gravy. Bowls were piled high with string beans, seasoned with bacon and onions and cooked with little potatoes. There were cabbages cooked with ham hocks and bowls of golden corn. Desert consisted of apple pie, peach cobbler, and six-layer cakes – wholesome food for hearty workers."

A "Combined Harvester" at work near Walla Walla, Washington, circa 1898. This machine included a header, thresher, separator, fanning mill and sacker. It could cut from 60 to 125 acres a day. Steam engines were sometimes used instead of horses.

COMBINES are harvesting machines that "combine" both harvesting, and threshing operations. The first practical "combined machine" was built in 1875, by a gentleman by the name of D. C. Matteson.

In the 1880's Benjamin Holt, a Stockton, California, farm implement dealer, began manufacturing a combined harvester and thresher, drawn by 24 horses, that was used in the dry wheat fields of the west for the next thirty years. The Holt Brother's firm evolved into the Caterpillar Company. (See Holt's story in the Steam Engine section.)

This interesting letter from a Fairfield, California, dealer appeared in an 1892 issue of *Farm Implement News*:

"The haying is over, the barley is now being cut, and wheat will soon follow. Several farmers have accumulated fortunes of $20,000 to $500,000 over the past thirty years, and while the land is not as rich as in former times, today's farmers know how to better cultivate it."

"Plows and cultivators have reached a high degree of perfection; but here we must make mention of one implement, the *combined harvester* and *thresher*. The cradle scythe was formerly used, followed by the *binder* and *header*, which in turn, were followed by the little horse-powered *separator*; then the large steam-powered separators. Now all of these are being superseded by a machine called a *harvester* which heads, threshes and sacks the grain in one operation for as little as $1.25 an acre. We now put grain in a sack for what it used to cost to put it in a stack."

"An ordinary sized machine cuts a swath sixteen feet wide, and requires twenty-four horses and four operators, and covers twenty-four acres a day."

"Within the past year two *traction steam engines* have been used for drawing plows and now one is going to draw a very large harvester which is expected to cut and sack forty acres per day."

Another, slightly later, description of a combine follows: "The combined harvester and thresher is a threshing machine with a harvesting mechanism at the side that conveys the headed grain from a wide swath directly to the thresher cylinder. The cutting cylinder and the elevating machinery is much like that of a harvester. The threshing mechanism is the usual type."

"These machines have an enormous capacity, harvesting up to 100 acres or 2,500 bushels of grain a day. The swath varies from 18 to 40 feet, and the power is furnished by horses, or a traction engine. From 24 to 36 horses or mules are under control of a pair of leaders driven by lines. Following the leaders there are usually two sets of four, the remainder of the animals are arranged in sets of six or eight. In this way one man is enabled to drive the entire team. At least three other men are required to operate the combine, one to have general supervision, one to tilt the cutter bar, and one to sew and dump the sacks when they accumulate in lots of six or eight. The largest machines are operated by steam traction engines."

Several firms offered tractor-pulled combines in the early 1920's, including the Holt Brothers, Avery & Sons, McCormick - Deering, Massey - Harris, Advance - Rumely, and John Deere, who claimed a savings of fifteen to twenty-five cents a bushel in harvesting costs over those of horse-drawn machines.

SILOS and ENSILAGE-CUTTERS. *Ensilage*, or *silage*, is an animal fodder made by cutting green corn stalks (with or without the ears), alfalfa, and other vegetable matter into short pieces that are preserved in a *silo*, which is an almost airtight storage tower. The word silo does not appear in any early American dictionary because it is of French origin, meaning "pit".

The first silos in America were large pits dug into the ground, Indian style. Elevated *cylindrical* structures probably evolved as farmers sought to add to the capacity of their existing silo pits by encircling them with wooden fences or rock walls. The first commercial silos were built in 1879, however, they were not widely accepted until the late 1890's.

The Department of Agriculture could only locate 99 farmers who owned silos in 1882, but within two decades a silo could be found on nearly every dairy farm in the country.

The primary purpose of the silo was to provide a winter source of green roughage for dairy cattle, however, farmers soon discovered that silage could be a valuable food additive to hasten the growth of many other animals.

University studies of the 1920's found that lambs fed silage as a major portion of their ration saved farmers $1.00 for every 100 pounds of mutton produced. Cows fed a diet of silage along with oats and cottonseed, delivered from ten to twenty percent more milk, and of a higher butterfat content than non-silage-fed animals.

In the earliest attempts to preserve fodder for winter, it was placed in pits or tanks, sealed with earth, and cooked with steam. Later it was discovered that it kept just as well when packed tightly into a silo; no cooking was needed.

Decay begins at once, and silage becomes very hot – as in a giant green compost heap. This decay uses up the air in the silo and changes it to carbon dioxide. Eventually the heat and lack of air stops the decay and the silage will keep for as long as no air can get to it. Only the top foot or two of fodder in the tower is exposed to very much air.

Silos were built from a variety of materials including stones, lumber, bricks, concrete blocks, cast concrete and clay tile. Many were constructed like barrels, with wooden staves and metal hoops. They ranged in size from ten to twenty feet in diameter, and from thirty to forty-six feet tall. The smallest size could feed a dozen cows for a period of 180 days. The largest size contained enough fodder to supply a herd of seventy cattle with food for 240 days.

Corn destined for silo storage was cut just after the milk stage of the kernels, while the leaves and stalks were still green and juicy. Neither corn nor alfalfa were allowed to lie around very long before being cut up and placed in a silo.

Prior to mechanization, specialized crews were often hired to fill silos. When tractors and portable agricultural engines became available, chain conveyors, or blowers and chutes, were used to elevate fodder to the top of the tower as it emerged from the cutter. Some creative farmers used their model-T Ford's elevated rear wheels as a power source.

Ensilage cutting machines of the 1890's ranged in size from hand-operated, lever-style choppers, to large rotary cutting machines which were run by two or four horsepower treadmills. Tractor-powered ensilage cutters and blowers of the 1920's and 1930's ran at speeds of 800 to 1,200 rpm.

CORN HUSKERS and SHREDDERS. In the South you *shucked* corn, while the same process in the North was called *husking*. Edward Hazen described a typical "husking bee" in his 1837 book, *Panorama*: "When corn has become ripe, the ears with the husks, and sometimes the stalks, are deposited in large heaps. To assist in stripping the husks from the ears, it is customary to call together the neighbors. In such cases, the owner of the corn provides for them a supper, together with some means of merriment and good cheer. This custom is most prevalent where the greater part of the labor is performed by slaves. The blacks, when assembled for a husking match, choose a captain, whose business it is to lead the song, while the rest join in chorus. Sometimes they divide the corn into two equal heaps, and apportion the hands accordingly, with a captain to each division. This is done to produce a contest for the most speedy execution of the task. Should the owner of the corn be sparing of his refreshments, his want of generosity is sure to be sung at every similar frolic in the neighborhood." After slavery was abolished white folks continued the tradition of husking bees.

Hand-husking later became assisted by the development of a special leather "husking glove" with one or two steel hooks mounted inside the palm. At least a dozen styles of gloves and "husking pegs" were patented from 1860 to 1880.

Corn husking actually became a competitive sport in the 1920's and 30's. Nearly 40,000 people attended the 1932 National Competition in Henry County, Illinois, where a left-handed husker set a new world record of 45 ears a minute.

HALL'S PATENT.

BRINKERHOFF'S PATENT.

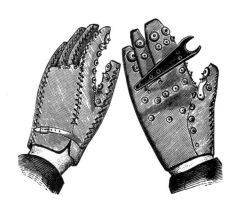

CORN SHELLERS. The first crude device used in the shelling of Indian corn was a simple horizontal iron bar fastened to the top of a box. Sometimes the edge of an inverted shovel served the same purpose, the dry ear of corn being rasped back and forth across it. Another method of separating the kernels from a cob was to employ a mallet to drive the ear through a fixture with a hole just large enough to admit the cob, leaving the detached kernels to fall into a box.

By the 1840's metal-toothed cylinder and disk-type corn shellers were available. They ranged from the small Clinton, Burral, and Harrison brands of early one-holers, to the huge steam-powered field machines of the 1890's. One of my dad's most vivid childhood memories was that of his father's purchase of a one-hole, hand-cranked cast-iron corn sheller in the 1920's. The South was still relatively poor and neighboring North Carolina farmers came from miles around to use my grandfather's new Sears & Roebuck corn sheller.

CORN BINDERS and CORN HARVESTERS. The earliest tool used for cutting corn was the common hoe. Later, sickles were used to top the corn, cutting off the stalk above the ear after fertilization had occurred.

One early method of building shocks of corn involved setting a center pole in the ground and inserting horizontal arms into it. Cornstalks were piled around the pole until a good-sized shock was formed. When the arms and pole were withdrawn the stalks were compressed and tied. A later method, used well into the 1900's, was to tie the tops of four hills, forming a saddle against which more cut stalks of corn was piled and bound in an upright shock.

Various forms of corn knives had been manufactured from an early date, including a style fastened to a man's boot, (which was not very practical). It was not until mechanical harvesters were developed that the full utilization of the entire corn crop, by a single implement, became possible.

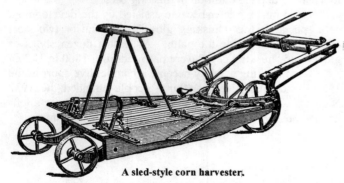

A sled-style corn harvester.

Many attempts were made after the 1870's introduction of the grain binder to build a *corn-harvesting* machine. The first horse-drawn stalk-cutters were sled-like implements, mounted on small wheels, which carried knives set at an angle on retractable cutting wings to cut the corn as it was grasped by two men riding on the platform. After each man had an armful, the cutter-sled was halted while both men stepped to the rear of the machine and set up their shocks. These cheap harvesters had a much larger capacity than hand cutting and could cover six or eight acres a day. They appeared in trade catalogs well into the 20th century (see page 172).

The D. M. Osborne Co. of Auburn, New York, came out with the first *mechanical corn harvester* in 1890, a horse-drawn two-wheeler which they exhibited at the Chicago World's Fair in 1893. It cut the standing corn and elevated it into a wagon drawn beside the machine.

McCormick vertical corn harvester, circa 1897.

The McCormick *corn binder* soon followed; the first model was pushed in front of a team of horses. In 1893, the Deering Harvester Company conducted a successful field trial with their new corn binder, and by 1900 there were several competing brands of *harvester-binders* available.

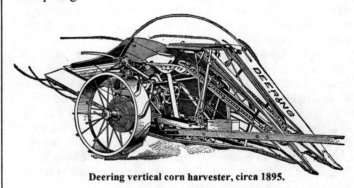

Deering vertical corn harvester, circa 1895.

Bound stalks were more easily handled and took up far less space than loose stalks gathered in the field and thrown in a wagon. A farmer using a corn-binding machine could cut and bind up to seven acres a day, while his neighbor with a corn knife had to work mighty hard to just bundle up an acre.

Still, more corn was husked from standing stalks in the field than were harvested in any other way. The stalks were commonly pasteurized during the winter; this being the cheapest method for the average farmer to feed his cattle.

CORN PICKERS were one of the last agricultural machines to achieve market success. In the early 1900's a growing scarcity of farm labor, and liberal wages paid for picking, encouraged many manufacturers to redouble their efforts to mechanize this labor-intensive process.

Since the 1880's various implement makers had thrown several hundred thousand dollars "down a well", trying to invent a corn picker that could *pluck-and-shuck* as fast as a skilled field hand. The first futile efforts had produced huge machines weighing up to four thousand pounds. In 1905 there were two basic machines on the market: the *corn picker* and the *corn husker*. By the 1920's the two machines had been successfully combined into one corn-picking implement that could snap the ears from stalks and husk them cleanly.

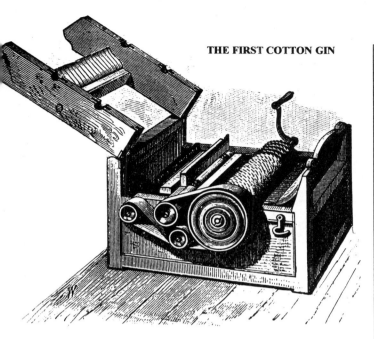

THE FIRST COTTON GIN

ELI WHITNEY invented one of the most important agricultural implements in the history of the United States, *a machine that picked seeds out of cotton.*

Born in Westborough, Massachusetts, in 1765, young Eli was recognized as a mechanical genius in a country where there were perhaps only a dozen so-called "mechanics", (today we would call them engineers). In his youth Eli Whitney was a self-employed nail-maker, hat pin-maker, and local blacksmith. By the age of twenty he had grown tired of making a living with his hands and decided to attend Yale University, but his father needed him on the farm and seven more years elapsed before Eli graduated. After receiving his degree in 1792, Whitney traveled south to Savannah, Georgia, where he read law while residing as a guest with the widow of the famous Revolutionary War general, Nathaniel Greene.

The economy of the state was generally depressed, there being no cash crops of any consequence. The only variety of cotton that grew well in the area was a worthless short-staple variety that required a full day's hand labor to remove two pounds of tiny seeds from three pounds of fresh picked cotton. Eli studied the motion of the slaves' hands as they did this tedious chore and within ten days he had constructed a machine that "did the work of fifty Negroes".

At first he used bent-wire hooks like those of a textile card, but much larger. These were placed in rows on a hand-cranked revolving cylinder, whose teeth passed through a frame of parallel wires that caught the cotton and drew it through the wire grate, leaving the tiny seeds on the other side. The fragile wire-teeth proved to be easily damaged, so Whitney replaced them with a series of thin circular saw blades shaped like bird's beaks. Behind the saw-cylinder a faster-turning reel of parallel brushes removed the cleaned lint from the blades and blew it out of the boxed enclosure.

Every part of this machine had been made by Whitney, who invited a group of local farmers over to view a demonstration of his new invention. The farmers saw how Whitney's gin differed from the old-fashioned roller gins that were currently in use in the Bahamas, and immediately recognized its great potential.

Before the week was out someone had broken into Whitney's workshop and made detailed sketches of his unpatented invention. Whitney and his partner (Mrs. Greene's fiancee) Phineas Miller, had limited funds and their patent application proceeded too slowly to suit several impatient planters who hired a local blacksmith to duplicate Whitney's gin. In fact, many of the farmers who had witnessed Eli's initial demonstration had already rushed to seed all their acreage in fast growing short-staple cotton.

The situation rapidly went from bad to worse. In 1794, when the partners were preparing several machines for lease, Whitney and his workmen fell ill from heat exhaustion. The few honest farmers, who were waiting to purchase their promised machines in time for the harvest, were infuriated by the partners' new lease proposal. They claimed that the proposed deal, (a pound of cotton for every three pounds processed), would have made Whitney and Miller instant millionaires. In 1795 Whitney's workshop mysteriously burned to the ground.

In a nutshell, the partners eventually went broke. They spent the entire $50,000.00 settlement, which they had received from various state legislatures, in an endless series of law suits against dozens of patent infringers. However, the introduction of Whitney's gin continued to act like magic on the production of cotton. In the eight year period from 1792 to 1800, cotton exports from the United States increased more than a hundred-fold. The tonnage rose from 138 thousand pounds to 18 million pounds and the value rose from $30,000 to $3,000,000.

Slavery, which had previously become unprofitable because of low-value crop yields, suddenly became viable again as a cheap labor source for cultivating the huge cotton plantations. In 1790 prime black field hands brought $200 at auction. By 1812 the median price for a healthy male slave between 18 and 25 years of age had risen to $650.

The average black field hand cost a planter less than $50 a year to house and feed, and could produce six bales (3,000 pounds.) of cotton annually. By the beginning of the Civil War there were four million slaves living in the South; in Mississippi they made up nearly half of the total population.

Southern plantation owners enjoyed one of the highest living standards in the world. Nearly every doctor, lawyer, speculator and investor, and those newly-married to rich widows, got into the cotton business. Former white laborers who could scrape together enough money for a mule, a plow and a sack of seed also took up the cultivation of "King Cotton". Prices continued to rise dramatically as English mills and Northern states turned the crop into finished goods as fast as they received it.

In 1791 the entire cotton production of the United States amounted to only sixty-four bales, but by 1834 output totaled 1,000,617 bales annually. In the 1850's new steam-powered cotton gins replaced horse-powered machines and production increased another hundred fold. Three Negro slaves, tending a steam-powered gin, could process from one to two tons of cotton a day.

Eli Whitney left the South and eventually bounced back financially. In 1804 he obtained a U. S. Government order for the production of ten thousand muskets.

THE "BOSS" AND "ACME" BARREL CHURNS

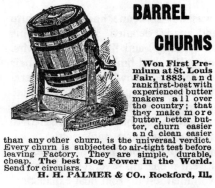

Won First Premium at St. Louis Fair, 1883, and rank first-best with experienced butter makers all over the country; that they make more butter, better butter, churn easier and clean easier than any other churn, is the universal verdict. Every churn is subjected to air-tight test before leaving Factory. They are simple, durable, cheap. The best Dog Power in the World. Send for circulars.

H. H. PALMER & CO., Rockford, Ill.

Delaware Co. Creamer

Requires no lifting or handling to skim or clean it. It is the prince of LABOR-SAVING Creamers. It will last for 20 years. It is warranted to do all we claim. To one man in every town where not already introduced we will make a special private offer. Address

Delaware Co Creamer Co. BENTON HARBOR, MICH.

"GET THE BEST."
HIGHEST AWARD at the CENTENNIAL.

THE IMPROVED BLANCHARD CHURN PRICE GREATLY REDUCED

Cheap, because so well made, durable, and efficient. Nine sizes made, churning from one, to 150 gallons. Warranted to be exactly as represented. Sold by all dealers in *really* first class Farm Machinery.

PORTER BLANCHARD'S SONS, CONCORD, N. H., Sole Manufacturers. Send for Circulars.

THE Fairlamb System OF Gathering Cream.

Send for Catalogue to

Davis & Rankin, SUCCESSORS TO Davis & Fairlamb, DEALERS IN Creamery Supplies.

24 to 28 Milwaukee Av. Chicago, Ill.

COW MILKER.

I have a MILKING MACHINE that will milk any cow in five minutes, clean. Its size is so small that it can be carried in the vest pocket. No more short teats or sore teats by bad milking; no more dirty or filthy milk. It can be used by a small boy; my boy of ten years milked to-day ten cows in less than one hour. I can supply the machine for $5 each, to those who wish to try it, or it can be seen here at any time.

WM. CROZIER, Northport, Long Island. N. Y.

THE O.K. CHURN Has Improvements over THE BEST!

Makes the Best Butter.

Easy to clean, easy to operate. Will not wear out; cover castings will not break. Send for circular.

JOHN S. CARTER, Sole manufacturer, SYRACUSE, N. Y.

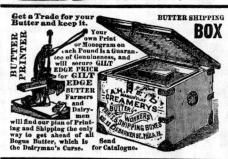

Get a Trade for your Butter and keep it.

BUTTER PRINTER

Your own Print or Monogram on each Pound is a Guarantee of Genuineness, and will secure GILT EDGE PRICE for GILT EDGE BUTTER Farmers and Dairymen

BUTTER SHIPPING BOX

will find our plan of Printing and Shipping the only way to get ahead of all Bogus Butter, which is the Dairyman's Curse.

Send for Catalogue.

SHIPPERS OF MILK, ATTENTION! WARREN MILK BOTTLES.

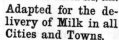

PATENTED MARCH 23d, 1880.

Adapted for the delivery of Milk in all Cities and Towns.

A Long Needed Want at last Supplied.

DESCRIPTIVE CIRCULARS ON APPLICATION

Warren Glass Works Co. A. A. 72 Murray St., NEW YORK.

Cow Milker.

SENT FOR $1.00 PER SET, PRE-PAID, BY MAIL.

THE NEW UNIVERSAL COW MILKER is warranted to MILK BETTER than any other. It is perfect in every particular. It holds or starts the milk at pleasure. No bulge, rubber, holes, to irritate the teats. Simple, durable, clean, and always ready. I. W. PARMENTER, box 464; 15 Murray St., "Methodist" Office, N. Y. City.

CHAMPION MILKER. Patented Aug. 6th, '78, Price $2, by mail. Agents wanted. Address J. W. GUERNSEY, Gen'l Ag't, 78 Cortlandt St., New York.

CHEESE FACTORY SUPPLIES.

Portable Engines and Boilers, Vats, Hoops, Presses, Screws, Cans. Every article needed in Cheese Factory or Creamery. Send for Circulars.

GARDNER B. WEEKS, Syracuse, N. Y.

"APPLE CORER AND SLICER," CHEAP, simple, and durable. Price 50c. each by mail. For State and County rights.
Address H. W. WILLIAMS & CO., Galesburg, Ill.

GOOD MEN WANTED

to sell the celebrated cow fetter. It sells on sight. Warranted to make the worst kicking cow gentle to milk, in three days. There is nothing equal to it for breaking heifers. Retail price $2. For further information send for illustrated circular to

H. J. SADLER, Sole Proprietor, Warren, Trumbull Co., Ohio.

DAIRY GOODS.

We make, from the best material, superior articles of Dairy Goods that are models of strength and simplicity. Rectangular Churns, Lever Butter Workers, Factory Churns and Power Workers. 2 gold and 14 silver medals awarded for superiority. One Churn at wholesale where we have no agent. Write for prices. All goods warranted.

CORNISH, CURTIS & GREENE, Fort Atkinson, Wis.

Pat. Channel Can Creamery.

SOMETHING NEW FOR SMALL DAIRIES. AUTOMATIC BUTTER-WORKER,

Just invented, without Gears or Cogs. We furnish Churns, etc. First order at wholesale, where we have no agents. Manufactured at Warren, Mass., and Fort Atkinson, Wis. Send for circular.

W. E. LINCOLN CO., Warren, Mass.

The Rectangular Churn.

Simple, efficient, and *always reliable*. No inside fixtures. Fifty per cent in labor saved over *any other churn*. Five sizes made. The Highest award given over all competitors at the late Dairy fair in Chicago. An energetic man wanted in every town, to act as agent. One churn sent at wholesale where we have no agent.

CORNISH & CURTIS, Fort Atkinson, Wis.

STAR CHURNS.

(See Cut),

ALSO,

Spain's Churns, 'Rapid' I. C. Freezers, Improved Tree-Tubs. Send for descriptive circulars and prices.

CLEMENT & DUNBAR, Philadelphia, Pa.

BUTTER WORKER.

The most effective, simple and convenient yet invented. Works 30 lbs. *in less than 5 minutes, thoroughly working out buttermilk and mixing the salt.* AGENTS WANTED. Send for circular.

A. H. REID, 6 N. Eighteenth St., Philadelphia Pa.

COOLEY SYSTEM FOR Butter and Cheese.

All the cream between milkings. No Skimming, no dust, no flies, no spoiled messes. Milk sweet for use after all the cream is off. Send stamp for circular.

Vt. Farm Machine Co., Bellows Falls, Vt.

LILLY'S BUTTER-WORKER

The cheapest and best Machine in the market; no hard labor required. Try it, and see for yourselves. Only $15 for a thirty pound machine that will take all the milk out with five minutes' work.

HENDERSON & CO., 316 Race St., Phila., Pa.

BUTTER COLOR.

After fair trial and severe tests it was awarded Centennial Prize Medal. WHY IT IS SUPERIOR TO ALL. 1st. It has no taste or smell, and is as harmless as water. 2d. It is liquid, is easy to handle, and is mixed in cream before churning. 3rd. It produces a color resembling June Grass Butter. 4th. It is the only article that will color the butter and not the buttermilk. 5th. It gathers all butter materials, increases the weight more than will pay for the color used. It is the *best* ever known. Send your address on postal card for my receipt book, *free*. It tells how to make butter, pack, preserve, extract rancidity.

MRS. B. SMITH, 327 Arch St., P. O. Box 1954, Phila., Pa.

SPAIN'S PATENT CHURNS.

Centennial Medal Awarded.

In use over 25 years.

Removable Dasher.

Made of White Cedar with galvanized hoops.

Send for circular and prices.

CLEMENT & DUNBAR, Philadelphia.

BUTTER CHURNS. Butter was the most profitable product a cow could produce, but first you had to squeeze all that watery liquid out of the cream and bind those molecules of pure butterfat. A noted author of the early 1800's estimated that 6,000 different versions of the butter churn had been invented since the ancient goatskin variety. The domestic market was flooded with churns of every description shortly after the Civil War. (The ads on the facing page date from 1878 to 1883.) Among the most collectable churns available today are the glass-bottomed Dazey churns that were mass-produced in the early 1900s for family use. The tiny number ten (1-quart size) is the rarest, having brought up to $1,000 at antique auctions.

Will make 2 lbs. of milk charged butter from 1 lb. butter and 1 pt. of milk, extra heavy clear glass jar, steel screw cap, heavily nickel plated, 1 piece cast iron frame, wood handle, retinned steel dash rod, white maple dashers. Perforated removable strainer in cap.

FAMILY GLASS CHURNS

COW MILKERS, or MILKING MACHINES, as they were later called, were slow to evolve beyond the concept stage. The first models, which appeared circa 1865 – 1870, promised to end one of the most irksome of all farm chores, and claimed: "Our new milking machine will fill a 14 quart pail in 7 minutes flat. The milk gathers no impurities, as in the usual handling of the teats and udder, in which numerous scales of skin, hairs, and other cow-flavored matter fall into the milk pail. The relief from the exertion of the wrists and arms is very welcome to the milker."

William Crozier, of Long Island, New York, introduced a Scottish-made milking machine to the American market in 1877, which he claimed had been used in a noted Scotch dairy for eight years. It consisted of four slender metal probes which were inserted into the teats. To these tubes of pure silver were attached four slightly larger rubber hoses that terminated in one spout just below the rim of an open milk pail. No suction apparatus was attached; apparently it was purely a gravity flow device. After milking, the manufacturer recommended that the tubes be soaked in a pail of cold water before being used again.

Pat'd May 28th, 1878, is just the thing every one needs who keeps cows. Any child can use it, while the men are at work in the field. Sent to any part of the U. S. on receipt of **$2.00**. Send for descriptive illustrated pamphlet, containing internal views of a cow's teats and bag, dissected by Drs. White and Wilson, of this city, and testimonials from Vet. Surgeons and Farmers, who have tested the milker for over six mos.

The editor of the *American Agriculturist* stated in 1878 that he had examined twenty different milking devices and that all had produced some degree of unfavorable results, some even drawing blood. Still, a market of at least a million sore-armed laborers was out there, begging for relief. These farmers' sons were rising at 5:30 every morning to spend several cold hours in the milk barn, only to have to repeat the same chore again that evening. The nation's 12 million milch cows had to be milked twice a day, seven days a week. It is said that this unending drudgery drove many an innocent country boy into bootlegging or an early military career.

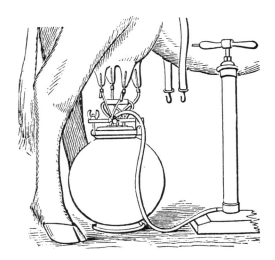

Temporary relief came in the 1880's when a vacuum hand-pump and a covered jar were added to the previously invented gravity-flow udder attachments. The Mehring foot-powered milker, invented in 1891, resembled a rowing machine and could milk two cows at once. However, twenty-five years elapsed before the first really practical milking machines evolved, circa 1916.

Single-cylinder gasoline engines powered a variety of successful vacuum pump milkers until electricity took over in the 1930s. In 1947 an International Harvester pamphlet stated: "Milkers have eliminated the hand milking chore on many American farms. Modern milking machines extract the milk from cows' udders with a gentle massaging action."

DAIRY GOODS

RAILROAD MILK CANS

"Iowa" Pattern — Heavy open hearth steel, double tinned, 6 in. 1 piece, double seamed locked neck, breast double seamed to body, bevel edge iron band, heavy iron bottom band, deep sanitary seamless covers, wired top and neck, patent round hollow steel handles. 1 in pkg.

Trade	Actual Wt.	Each
T7755— 5 gal.	5 gal.12 lbs.	$2.85
T7758— 8 "	8 " 15 "	3.55
T7760—10 "	10 " 18 ".	3.75

"Hercules"—Heavy open hearth steel, heavily coated with pure tin, double seamed, locked neck, patent bevel edge breast band, 1 piece round handles, 1 piece patented sanitary cover with double rim, bottom and heavy hoop are combined in one indestructible bottom. 1 in pkg.

CREAM CANS

IX tin plate, without gauge, patented watertight cover forms air chamber on the outside below wire edge, no locking required, 1½ in. deep cover, extra long riveted ears, coppered wire bail, black enameled wood handle. Wrapped ½ doz. in crate.

Trade	Act.	Crated	Doz
T7410—10 qt.	9½	20 lbs.	$3.95
T7414—14 "	13	28 "	4.25
T7420—20 "	19	30 "	5.60

WELL MILK COOLERS

T7316 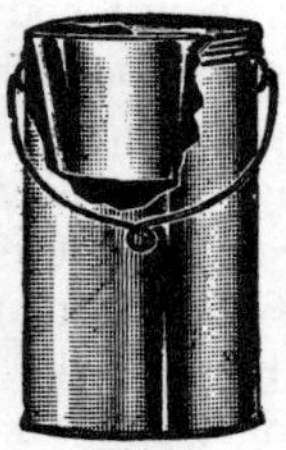   T7196

T7316—3 sizes, ⅙ doz. each of 4, 6 and 8 qt., heavy tin, large zinc screw top, will not rust, heavy wire bail. Asstd. ½ doz. in crate.
Doz $4.20

T7196—6 qt., 2 lbs., inside butter pan, heavy bright IC tin, 6 in. screw top, wire bail with loop for line. ½ doz. in crate

---

MILK STRAINERS

High Shape—Bright tin, IC plate, footed, brass twilled wire, double seamed in bottom, wire strainer. 1 doz. in crate, about 7½ lbs.

Diam.	Doz
T7159—9½ in.	$1.25
T7160—10 "	1.75

Low Shape—10 in., bright tin, IC plate, seamless, footed, 60 mesh tinned steel wire, double seamed bottom.
T5421—1 doz. in pkg. Doz $1.30

Milk Can—9½x5 in., bright tin, IC plate, detachable steel wire for cloth, 2½ in. brass strainer cloth.
T7162—1 doz. in crate, 8 lbs.
Doz $1.95

Milk Can—8¾x6, bright tin, IC plate, double seamed, footed, brass wire strainer.
T7163—1 doz. in crate, 10 lbs.
Doz $2.90

Milk Can—10½ in., 1 piece, IX plate, seamless body, 4½ in. brass wire strainer cloth, double seamed in bottom.

FLARING PAILS

Bright tin, IC plate, double seamed, well soldered, wire top, patent bottom, riveted ears, wire bail. 1 doz. in pkg.

WITHOUT WOOD HANDLE—
Trade	Act.	Doz
T7036—6 qt.	4¾ qt.	$1.65

WITH WOOD HANDLE—
Trade	Actual	Doz
T7038— 8 qt.	6 qt.	$1.85
T7040—10 "	9 "	2.00

IXL TIN DAIRY PAILS

Tin plate, double seamed body, riveted ears, coppered wire bail, natural finish wood handle. 1 doz. in crate.

Trade	Actual	Doz
T7026—10 qt.	10 qt.	$2.90
T7027—12 "	12 "	3.25
T7028—14 "	14 "	3.65

T7140—Trade 10 qt., actual 9 qt., flaring, bright tin, IC plate, swaged body, double seamed, back handle, fine brass strainer, wire bail, black enameled wood handle. ½ doz. in crate, 14 lbs. Doz $4.20

"STEEL-CLAD" DAIRY PAILS

Bright Tin, IX Plate—Double seamed and soldered, wired and reinforced, strongly riveted ears, heavy coppered wire bail, natural finish wood handle, each labeled. ½ doz. in crate.

Trade	Actual	Doz
T7046—10 qt.	10¼ qt.	$3.85
T7047—12 "	12½ "	4.10
T7048—14 "	14½ "	4.50
T7049—3 sizes, ⅙ doz. each 10, 12 and 14 qts., (trade sizes). 1 doz. in crate.		Doz $4.10

Lipped

Bright Tin, IX Plate—Double seamed and soldered, wire top rim with lip, strongly riveted ears, heavy copper wire bail, black enameled wood handle, each labeled. ½ doz. in crate.

Trade	Actual	Doz
T7056—10 qt.	10½ qt.	$4.50
T7057—12 "	12½ "	4.80
T7058—14 "	14½ "	5.10
T7059—3 sizes, ⅙ doz. each 10, 12 and 14 qt. (trade sizes). 1 doz. in crate.		Doz $4.75

IXX Coke Plate—Bright finish. No. 4 body wire, ⅝ in. raised bottom with double seam, extra strong throughout, heavy tinned riveted ears with reinforcing plate. No. 0 tinned wire bails, with black enameled wood handles, tested and guaranteed. ½ doz. in crate.

Trade	Actual	Doz
T7050—10 qt.	10½ qt.	$4.60
T7051—12 "	12½ "	5.10
T7052—14 "	14½ "	5.50
T7053—3 sizes, ⅙ doz. each 10, 12 and 14 qt. (trade sizes). Asstd. 1 doz. in crate.		Doz $4.95

STRAINER MILK PAILS

T7062—Steel-Clad "Jersey," trade 12 qt., actual 12½ qt., IXXX coke plate highly polished, double seamed, square lip, soldered back handle, heavy malleable iron ears, heavy wire bail, strongly riveted, black enameled wood handle. ¼ doz. in crate, 17 lbs. Doz $8.50

T7412—"Gem," trade 12 qt., actual 12 qt., IX charcoal tin, double seamed, back handle, detachable strainer, heavy malleable iron ears, wire bail, black enameled wood handle. ½ doz. in crate. Doz $8.90

DeLaval Separators

- Beautiful gold and black finish.
- Completely enclosed gears.
- Improved regulating cover.
- New turnable supply can.
- Easier starting and turning.
- New oil window.
- Wonderful floating bowl.

De Laval Golden 50th Series

CREAM SEPARATORS were not really an American invention. After years of experimenting, German engineers came up with a machine in the 1870's, that separated cream from whole milk. The Lefeldt separator arrived on the market in 1877 and Dr. De Laval followed in 1878 with his highly successful machine. The Iowa Dairy Separator Company began production of one of the first American-made cream separators in the early 1880s. It was designed around a series of internal disks, as were the more than two dozen other domestic-made machines that followed before WWI. The one exception was the popular Sharples brand separator, a tubular design, dating back to the 1880s.

The market for this newfangled method of getting the cream out of milk was huge – so huge in fact that Sears Roebuck devoted the first seven pages of their 1908 catalog to a dramatic sales pitch that was hard to ignore. Sears' $26.30 price tag for a 250 lb. capacity machine was $44.00 cheaper than most competitors.

Sears reminded farmers that they spent a third of their workday feeding, herding and milking cows, and the only real money they made was on the butter. And if a third of the butter was lost through improper cream recovery, the farmer was not making any profit at all.

Before the widespread use of separators, there were a number of methods employed to remove the butterfat from whole milk. The oldest way was to set the milk in crocks or tin pans in a cool place and wait for the cream to rise to the surface. Another later method was to set the milk in tall, narrow cans submerged in cool running water, or a vat of ice water. The third way was to separate the cream, or a portion of it, by diluting the whole milk with a large quantity of cold water. Many farmers simply loaded their wagons with cans of warm milk and drove to a nearby creamery where they had it skimmed by power separators. They sold the butterfat to the creamery and then hauled the skimmed milk back to the farm, where it was fed to chickens, calves, pigs and other livestock as a highly nutritional food, rich in sugar, protein and other bone and muscle building ingredients.

At the peak of production, in 1918, nearly 200,000 hand-cranked cream separators a year were coming off the production lines of thirty American manufacturers. By the 1920's nearly every farm wife owned a shiny new cream separator. She used the cream to make butter for home use and sold her surplus to the local storekeeper. *Farm-fresh* butter was a valuable commodity. Most factory-made butter was stale or otherwise of an inferior grade.

50TH GOLDEN ANNIVERSARY De Laval First in 1878 Best in 1928

HOLSTEIN CATTLE.

First Prize Herd at N. Y. State Fair. 1879, 1881, 1882, 1883.

**Largest Herd,
Best Quality,
Most Noted Families.**

At head of herd are the four best bred milk bulls living.

We now offer for sale the choicest lot of yearling bulls and heifers ever collected in one herd, as their pedigrees show, and all backed by wonderful records.

Fine **CLYDESDALE** and **HAMBLETONIAN STALLINNS**, at low figures.

Catalogues on application. Correspondence and personal inspection solicited.

SMITHS & POWELL,
Lakeside Stock Farm, Syracuse, N. Y.
Mention the Farm, Field and Fireside.

ISAIAH DILLON AND SONS. LEVI DILLON AND SONS.

IMPORTERS AND BREEDERS OF

NORMAN HORSES

(Formerly of firm of E. Dillon & Co.)

THREE IMPORTATIONS IN 1883.
200 Head of Normans on Hand.

STABLES AND HEADQUARTERS LOCATED AT NORMAL,

Opposite the Illinois Central and Chicago & Alton Depots. Street cars run from the Lake Erie & Western, and Indianapolis, Bloomington and Western Depots, in Bloomington, direct to our stables in Normal. Address.

DILLON BROS., NORMAL, ILL.

BROOKBANK HOLSTEINS

FOR SALE.

15 CHOICE YEARLING BULLS, COWS, HEIFERS and CALVES; also 5 CHOICE TWO-YEAR-OLD HEIFERS, bred to **Mercedes Prince (2150)**, the only son of the cow Mercedes (723) and the renowned Holstein bull Jaap (452), who headed the **Champion herd of 1883**. First at Minneapolis, Des Moines, Chicago, and St. Louis; also winning the Grand Daisy Herd prize, at Minneapolis, and the Grand Herd prize of the Illinois State Fair for milk breeds. The average milk and butter records of the dam and grandam of **Mercedes Prince (2150)** are the largest yet recorded, viz.: 81 lbs. 12 oz. milk in one day; 21 lbs. 11 oz. butter in 7 days, placing him at the **head of all Dairy bulls by actual tests.**

THOMAS B. WALES, Jr., Iowa City, Iowa.

ON HAND, APRIL 1st, 1884,
AT OAKLAWN FARM,

50 Imported Stallions,

Weight 1,500 to 2,300 lbs., well acclimated and ready for service. Also

**100 YOUNGER STALLIONS
and
125 IMPORTED MARES.**

Nearly all the above registered in the

**PERCHERON STUD BOOK
OF FRANCE,**

which is the only draft horse record of that country.

Notwithstanding this immense stock, my importations for 1884 have already begun. The first installment of

20 FINE LARGE STALLIONS

will be shipped from France the first week in April, to be followed by

HUNDREDS OF OTHERS

during the season.

ALL STALLIONS GUARANTEED BREEDERS.
Catalogue free. Address

M. W. DUNHAM,
Wayne, Du Page County, Illinois.
35 miles west of Chicago, on C. & N. W. Ry.
Mention the Farm, Field and Fireside.

A Remarkable Jersey Cow.

We have been favored by Mr. Harvey Newton, of Southville, Mass., with a record of his Jersey cow "Abbie," from April 17th, 1876, to March, 1877, during which time she yielded 10,070 lbs. of milk, as follows: April 17th to 30th, 417 lbs.; May, 1,365 lbs.; June, 1,406 lbs.; July, 1,247 lbs.; Aug., 1,155 lbs.; Sept., 992 lbs.; Oct., 907 lbs.; Nov., 794 lbs.; Dec., 788 lbs.; Jan., 1,877, 707 lbs.; Feb., 551 lbs.; March, 371 lbs. Total, 10,700 lbs. On April 15, 1877, she calved again. The butter produced within the year was 486 lbs.; besides which, milk and cream were supplied for family use. A portrait engraved from a photograph of this excellent cow is given on the preceding page.

The Patent Self-Acting Cow Milker M'f'g Co.

Patented May 28th, 1878.

Every one who owns a cow should have one of our wonderful Milkers. A child can use them. Sent free to any part of the United States on receipt of $2. Send for our Illustrated Pamphlet on the Cow, containing sectional views of a cow's teats and bag dissected and scientifically explained, by Drs. White and Wilson of this city. Sent free to any address.

GEO. KING, President.

Office, 575 Broadway, New York.

The Centennial Medal and highest Diploma of Merit, was awarded, at the Philadelphia Exhibition of 1876, to

Bullard's Oscillating Churn.

Col. Waring, of the *American Agriculturist:* "I never have seen a churn for which I would exchange it." A. W. Cheever, of New England Farmer: "The ease with which it works has lessened the dread of churning day ONE-HALF." L. B. Arnold, of New York. "Thinks the work done better than in the dash churn." Hon. Harris Lewis, New York: "It is the best Churn I ever used." For Descriptive Circular address,

JOHN T. ELLSWORTH, Barre, Mass.

FEED CUTTERS have been utilized by farmers since the iron age, when a heavy knife was probably employed to cut up dry feed for livestock. Later a box-style cutter evolved, with a knife hinged on the end that could shear off the fodder to any desired length. Rotary spiral cutters that revolved on a horizontal shaft were invented in the 1820's, and another machine, with sharp bladed spokes mounted inside a flywheel, was developed about the same time. These early implements were employed to cut corn, hay or straw for immediate use.

Stage line managers claimed a substantial savings in feeding their animals cut straw, rather than loose hay, and older horses were said to have their working lives extended by the more easily digested fodder. By the 1920's tractor-driven ensilage cutters, with attached blowers, were developed to fill silos with green and partially cured succulent crops for winter use on dairy farms.

FEED GRINDERS, or MILLS (i.e., mortar and pestles) were one of mankind's first agricultural inventions. Feeding the more easily digested cracked or ground grain to domestic animals, from chickens to cows, has been practiced for centuries, but only in the last hundred years has it become generally widespread; mainly because of the invention of portable iron mills. By the early 1900's nearly every farmer owned some form of a *feed grinder*.

The first *grain mills* were equipped with huge grinding stones. These heavy millstones had grooves chiseled in them to carry the grain to the outer edge between the milling surfaces. Most of the millstones quarried in the United States came from Esopus, New York, and were used in large windmills and also in waterwheel-powered gristmills that developed up to forty horsepower.

Metal milling plates began to replace stone wheels in the late 1700's and Napoleon's army carried portable mills with *metallic buhrs* during their expeditions. Nearly all of the plates used on early farm mills or feed grinders were made of chilled iron; the usual form being two flat corrugated disks. Cone-shaped buhrs were developed later to increase capacity by exposing a larger surface area. At first, metal mills were used primarily for animal feed, but stone ground flour remained popular for human consumption and is still preferred today by many health conscious consumers.

Horse-powered, cone-shaped, *sweep mills* were used to grind ear corn in the late 1800's. The usual arrangement was to hitch a team of horses to the end of the single sweep and walk them in a circle to power the mill. *Power mills* were operated from a pulley run from a horse treadmill, or by power supplied from steam, gas, water, or the wind.

Corn crushers were widely utilized for cattle feed at the turn of the century. They could be set up to chop the husks along with the ears, thus producing a finely ground mixture of cobs, kernels and husks at a rate of up to 50 bushels an hour.

Tractor-powered *hammer mills* and *roughage mills* with swinging hammers, mounted on swiftly revolving disks, generally replaced *burr mills* in the 1930's. The resulting cracked, crushed, and ground roughage was blown through a pipe into a storage bin.

BALDWIN'S AMERICAN FODDER CUTTER.

O. K.

Independent Feed Grinder.

Made entirely of iron, with heavy balance wheel and adjustable feed. A four to ten horse machine. Capacity from 20 to 40 bushels per hour. Can be run by tumbling rod or belt. Furnished with sacking elevator when desired. An immensely popular machine. Send for grinder Circulars.

If you have not seen our new Corn Sheller and Horse Power catalogue we should be pleased to send you a copy.

Sandwich, Illinois

POTATO PLANTERS. The Irish, or white potato originated in the Andes, where it was cultivated by the Incas. Spanish explorers brought the starchy tuber from Peru to Spain in the 1500's, and it spread throughout Europe, and then on to North America in the 1600's.

Our English cousins were the first to raise potatoes on a large scale. Potatoes became a major food source for poverty stricken Ireland during the 18th century. A catastrophic crop failure of 1845 – 46 caused a famine that nearly wiped out a whole generation of freckle-faced, redheaded, people, and caused a mass emigration of Irish families to America.

Mechanical potato planters, such as the "True's Potato Planter" (illustrated above) were available in 1870, but did not become widely used in the United States until the late 1890's. By 1910 at least three dozen firms had placed horse-drawn models on the market. Prior to the introduction of mechanical planters and digging machines, potatoes had not been a profitable crop for most American farmers.

Early models, such as those below, simply dropped and covered the bulky seeds in a somewhat haphazard fashion. But by the 1920's potato planters had become "almost human" in their chore of picking out a piece of irregular-shaped seed, dropping it in its proper place, and then covering it at the correct depth.

"Picker arms, revolving on the main axle of the John Deere *one-row potato planter,* pass through concave spaces in a seed-holding picker chamber. Each picker arm is equipped with two sharp picking points that pick out one piece of seed. As the arm passes downward the seed is forced off the picker points by contact with a bumper plate. Then the seed drops down a chute, into a trench made by a furrow opener, and is covered with soil by a disc at the rear of the machine. A rotary feed wheel, controlled by an automatic seed control, drops the seed from the hopper into the picking chamber."

POTATO DIGGERS were first developed in England, and in an 1888 contest sponsored there by the Royal Agricultural Society, there were two classes of machines: One "having shares to lift the potatoes, revolving prongs to separate them from the soil, and a screen to confine them within a given area." The other class consisted of the "old plow-type potato-raiser, with prongs behind the share to lift the tubers from the loosened earth." (British and Australian machines won all the prizes at this event.)

By 1892 the worm had turned; an article in *Farm Implement News* recorded the change: "In the United States, lifting potatoes with a plow has been practiced from time immemorial and potato-raising plows have long been manufactured and in common use. Three brands of plows made here have *recently* attained greater perfection than those produced by the English:

(1) The Shaker potato digger, manufactured by Deere & Co., is a raising-plow with a perfectly flat blade that enters the hill under the tubers, and allows them to slide up the blade and fall on the rods or fingers hinged on its trailing edge. A flange wheel in the rear lifts these fingers as each flange comes up and lets them drop back to a horizontal position. This action causes a continual shaking of the potatoes, knocking off the dirt, leaving the potatoes free to roll off the plow on to the top of the ground. A weed-fender on the front separates the vines and prepares the way for the shovel.

(2) The Boss potato-digging machine, manufactured by Rawson & Thatcher, of Corning, New York, since 1885, has an excellent reputation. It has a cart-like set of wheels with a tongue that projects forward from the axle, on which a gear is mounted that drives a multi-pronged separating wheel. As the machine moves forward, the vine tops, dirt and potatoes fall upon the revolving horizontal wheel-arms which separate the potatoes from the debris and leave them lying at one side of the course. One man and a span of horses can dig five acres a day with this time tested digging machine.

(3) One of our most successful potato digging machines is made by the Hoover & Prout, Avery Co. The Hoover digger has a huge chute-like forward blade mounted between two heavy iron wheels, with a driver's seat atop the axle. The forward projecting shovel passes under the hills, raising potato plants, soil and all, which are carried upward by a slotted elevator that shakes out the debris and delivers the cleaned potatoes in a narrow row behind the implement."

We should also note here that the Aspinwall Mfg. Company, of Jackson, Michigan, had become the world's oldest and largest maker of potato implements at the turn of the century. The two models shown (below, right) were made in 1878 and 1910. The one at left is European, circa 1865.

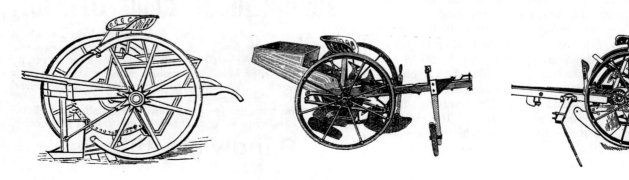

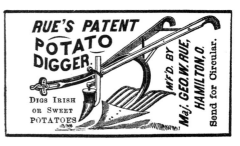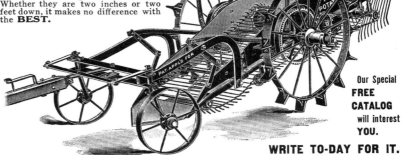

U. S. WIND ENGINE & PUMP CO., KANSAS CITY, MO.

THE HALLADAY STANDARD WIND MILL.

AS USED FOR

PUMPING WATER FOR RAILROADS.

We have had thirty years' experience in the manufacture of Railway Water Supply Material, such as Halladay Standard Wind Mills, Curtis Double-Acting Pumps, Horse Powers with Pumping Attachments, Hand Pumps, Steam Pumps and Boilers, Tanks, Outlet Valves, Spouts and Fixtures, and are prepared to contract for the erection complete, or furnishing of material for Wind, Steam, Horse or Hand Power Water Stations.

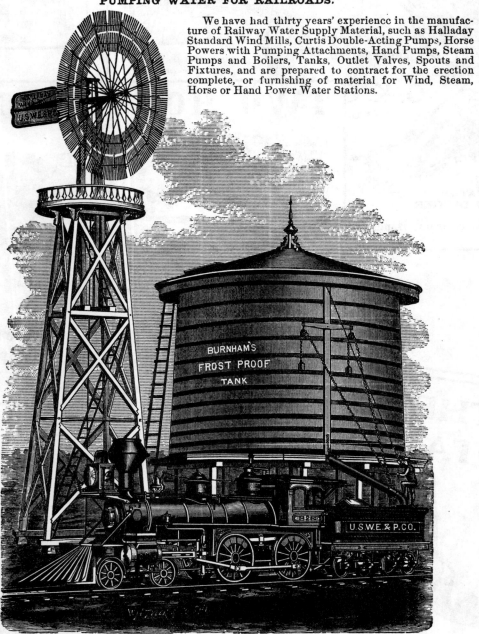

A RELIABLE AND ECONOMICAL WATER STATION.

In locations favorable for wind power we will guaranteee to fully supply from **4 to 30** engines per day when ample storage room is provided for water, to guard against occasional calm days. New roads can have a trial of the mill, if desired, and any company can pay by monthly installments of no more per month than the present cost to supply the station by steam, horse or hand power.

WINDMILLS have been around in one form or another for at least a thousand years. In Europe windmills were primarily used to power gristmills, but in Holland the Dutch built water-pumping windmills to literally stay alive. In November of 1421, the sea poured through Holland's dikes with a terrible roar and in only a few minutes drowned ten thousand souls in sixty-five hamlets.

When the Dutch settled New Amsterdam on the tip of Manhattan Island, in 1624, they began to build their traditional windmills up and down Broadway. These mills of stone, brick or wood were built as tall as four stories and had canvas sails from twenty to forty feet in length. They were kept constantly busy sawing wood and grinding grain for the bustling new city.

Sir George Yardley, Virginia's colonial governor, built our country's first post-mounted gristmill in 1618, on a one thousand-acre plantation located about twenty miles from Jamestown. William Robertson built a large post-mounted gristmill at Colonial Williamsburg, in the year 1720. Like governor Yardley's mill, built around a post, the entire mill house could be rotated 360 degrees in order to face the wind coming from any direction.

These first post mills were huge ponderous affairs that required their owners to go out and swing the entire building around when the direction of the wind changed. Dutch-style *smock mills* soon followed on which only the roof-cap section was rotated to redirect the sails. Some of the more advanced smock mills had a tiny windmill mounted on the rudder that swung the main wheel into the wind.

Windmill design did not evolve much over the next century; they still looked like giant inverted 3-story ice-cream cones complete with doors and windows. However, the western movement dictated new kinds of cheaper, lighter, machines that could be left alone to run by themselves.

Some pioneers constructed ground-level, paddle wheel-style windmills, mounted in boxes that faced the prevailing breeze. Others improvised canvas-sailed windmills mounted on twenty-foot tall wooden towers. The direction of the pivoting fan was controlled by a long diagonal beam that stretched from the top of the tower to a wheel or a stake on the ground.

In Kansas and Nebraska, cash-strapped farmers built "Battle-axe" windmills, using less than two dollars worth of lumber. A three-story tower was constructed using four cross-braced poles, and the X-shaped wheel had slats from old boxes nailed *battle-axe-style* to its arms. One of these crudely shaped contraptions, only ten feet in diameter, could pump enough water to support 75 head of cattle or irrigate ten acres of summer vegetables. In the wintertime these mills could be converted to saw firewood.

Things began to change in 1855 when an Iowa farmer hoisted a seventeen-foot diameter wooden-bladed turbine to the top of a 70-foot tall tower and began pumping several hundred gallons of water a day. Other Midwestern farmers soon realized the potential of this improved, multi-bladed, power source and nearly a thousand windmill patents were registered over the next twenty years.

One of the most popular designs of the late 1800's was a wooden-bladed windmill invented by a young Connecticut mechanic named Daniel Halladay. The big Dutch smock-style windmills had no speed governors, other than a primitive friction clutch or the removal of cloth from their wood-framed sails, and were easily damaged by sudden storms. Halladay solved this problem by using centrifugal force to close his wheel's jointed wooden fan blades, much like a flower folds its petals forward at night. The "Halladay Standard" windmill was manufactured in Batavia, Illinois, by the United States Wind Engine and Pump Company from 1865 until the stock market crash of 1929. Their first customers were the railroads, who were in a big hurry to lay tracks to Chicago. The huge water tanks, located every few miles along the railroad tracks, required extra large windmill pumps and Halladay was chosen to supply them.
(The 1885 engraving on opposite page is from their catalog).

Meanwhile, in Wisconsin, a mechanically inclined preacher of the gospel had also begun to experiment with different sizes and shapes of wood-bladed fans. In 1868, after several years of trial and error, Rev. Leonard R. Walsh, and his son, began manufacturing their new windmill design.. The "Eclipse" had a rudder to keep it facing into the wind, but it also had a smaller side-mounted vane that would turn the main wheel edgeways, out of the wind, during a storm. A weighted lever, attached to this side vane, would swing the fan blades back into the breeze when the storm was over. Both Halladay and Wheeler won gold medals for their new windmills at the 1876 Centennial Exposition.

In 1879 several dozen manufacturers, employing six hundred men, were selling over a million dollars worth of widely assorted windmill designs. By 1889 windmill sales had doubled, and in ten years they doubled again. In 1919, nearly two thousand men were working in American windmill factories, with total annual revenue reaching $9,933,00.

A few of the colorful brand names of the period were: Ace, Aermotor, Buckeye, Badger, Challenge, Dandy, Daisy, Decorah, Enterprise, Fountain, Monitor, Red Cross, Royal Crown, Sampson, Stover, Sunflower, Wonder, Princess, and The Steel Queen. At the turn of the century there were nearly one hundred different windmill makers doing business in the United States, including the catalog giants, Montgomery Ward, and Sears, Roebuck and Co.

The first modern scientific study of windmill design was conducted in 1888 by Thomas Perry, an engineer with the United States Wind Engine and Pump Company. Perry constructed a huge steam-powered centrifuge on which he mounted sixty-one different wheel and blade combinations, during 5,000 separate tests. The winning design, dubbed the "Aermotor", featured slightly curved, thin steel blades that turned in the slightest breeze. Perry geared down the wheel by a 3 to 1 ratio to increase pumping power and prevent the machine from flying apart during high wind conditions. The final result was a product that revolutionized the industry and its design remained virtually unchanged over the next one hundred years

Aermotor's catalog of 1895 claimed "Our 12-foot model will, in a fair wind, easily do the work of two horses, and will satisfactorily drive any machine that can be operated by two horses. It pumps water, grinds grain, cuts feed, shells corn, saws wood and drives machinery of all kinds."

E. C. MURPHY authored a consumer's guide to windmill performance in 1901 that put an end to exaggerated claims and forced many manufacturers out of business. With so many designs on the market the public was confused and the U. S. Government Printing Office gladly footed the bill for Murphy's report. Here are some of the numbers: Halladay's 30-ft. diameter mill, with 144 wooden sails, produced 1.07 horse power in winds blowing between 12 and 25 miles per hour. The 22-ft. "Eclipse" only mustered 0.18 h.p. The popular "Aermotor" with a 16-ft. diameter metal blade span, developed 1.53 horse power in a 20 mile an hour wind and could pump water at more than 100 gallons an hour.

Maintenance on the newer galvanized steel models was minimal, the gearbox oil only needed changing once a year. The last major advance in this technology came in 1933, when the main shaft on the Aermotor windmill was enlarged and the wheel arms were designed to screw into the hub for greater strength. Aermotor, now based in South America, is the world's largest windmill manufacturer, and some U. S. dealers can still supply parts for any model made prior to 1933. Under normal conditions these durable machines have a life expectancy of seventy years.

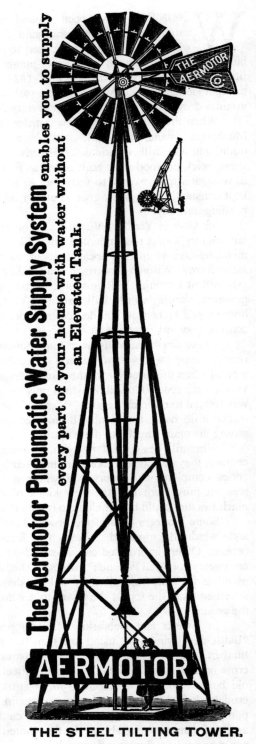

This is the only Windmill Tower which can be set close up against the house, directly in front of the door and right over the walk without being in the least in the way. It gives
HEAD ROOM
PUMP ROOM
TANK ROOM

There is nothing like it. It is a marvel of strength.

THE STEEL TILTING TOWER.

2,288 sold in '89
6,268 sold in '90
20,049 sold in '91
60,000 will be sold in '92

A complete STEEL WINDMILL and complete STEEL TOWER every 3 minutes during the working day. These figures tell the story.

We Were Pleased to Read Some Remarks

Made by the Aermotor Company, in *Farm Tools*, April 26. After publishing the first letter of Mr. Swartz, the Aermotor Co. proceed to say:

"This reminds one of the young doctor, who, on returning from his first important call, admitted that he had lost the mother and child, but declared that by the grace of God he hoped to pull the father through yet. It was the only testimonial regarding the storm presented from the great state of New York, where there was not sufficient storm to warrant any one in noticing it. We ask the attention of the implement dealers of the United States to the above statement of facts. Contrast the confident and positive tone of Mr. Swartz's letter of March 26, in which he states that 'all the Challenge steel towers in this region, save one, were blown down, etc.,' with the hesitating and unsatisfactory manner in which he attempts to modify and explain his slander in his letter of April 21."

We now wish to let these gentlemen (the Aermotor Co. and the U. S. Wind Engine Co.) settle the matter of the number of mills and towers belonging to each that were blown down and the fact of whether there was anything approaching a storm in the great state of New York. We wish to put on record our deliberate statement that out of all the hundred steel mills and towers in the state of New York, of our make, not one was injured or 1 cent paid out by us for repairs of any kind.

We have in our possession hundreds of testimonials from all parts of the country which we will be happy to exhibit. We print a few, please read carefully.

TESTIMONIALS.

H. F. Lehman, Good Thunder, Minn., gives a list of thirteen mills blown down within six miles of Good Thunder, and says: "There are about twenty-five to thirty Challenge wheels, O. K. and S. wheels within six miles around here, but not one damaged that I know of."

W. B. Brant, Renville, Minn.: "Yours of the 28th is at hand, and in regard to the wind mills, yours stood the storm the best of all. You cannot hear nothing from the farmers but 'the Challenge Wind Mill for me' up here since the storm."

C. E. Zeiglar, Blue Earth City, Minn., giving a list of twenty-one mills of various makes blown down in one county. No Challenge mills down. One hundred and thirty-five in use.

L. H. Smith, Lake Wilson, Minn.: "In reply to your favor of March 28, will say there were none of the Challenge wind mills either blown down or off the tower. Over fifty Challenge mills in this territory."

K. O. Lastrum, Elbow Lake, Minn.: "In reply to yours of March 28, will say that there was quite a large number of wind mills damaged in this vicinity, but as yet I have failed to hear of the first one of your make that has left the tower or been damaged in the least." Then follows a list of a number of mills that were blown down.

BROKEN BOW, Neb., April 2, 1892.

Challenge Wind Mill & Feed Mill Company:

Gentlemen:—In answer to your enquiry as to the damage done by the late storms, will say that of over 1,000 of your mills erected in this county not one has been reported damaged in the least, although scores of other mills have been blown down and broken to pieces; and as we have more than double as many mills as all other wind mills combined we think this is a good recommend. I have handled mills for years and have erected nearly every kind of wind mill now on the market, so that I am practically posted in regard to mills, and must say that in my judgment there is nothing that compares with your make of goods.

Yours truly, CORWIN JOHNSTON.

FERGUS FALLS, Minn., March 31, 1892.

Challenge Wind Mill & Feed Mill Company, Batavia, Ill.:

Gents:—Your circular letter of the 28th inst. received and noted. I shall furnish you a complete list of all wind mills and towers damaged in this vicinity in a few days, the names of the mill and also the names of the owners. All the Challenge mills that I have put up for the two years since I took hold of the Challenge mill, not a single mill has been damaged in the March storm. This is more than any other wind mill man in this city can say, for they all hang their heads down when one speaks about the wind mills, particularly the Challenge mill. Let them lie all they want, we are on top of the heap this time, and we will stay there in the future. Yours truly, M. FRANKOVIZ.

The above letter is from Mr. Frankoviz, and he has sent us a list of names of twenty-seven mills blown down close to his city and which list includes all the well-known mills made in this country to-day, and as we have more Challenge mills in that section than all other companies combined, we think it simply a fair sample of the merits of the mill.

WE ARE NOW IN A POSITION TO FILL ORDERS PROMPTLY, HAVING EQUIPPED OUR PLANT WITH ELECTRIC LIGHTS THROUGHOUT, WHICH ENABLES US TO RUN A FULL FORCE OF MEN NIGHT AND DAY, WHICH WE HAVE BEEN OBLIGED TO DO FOR MONTHS TO ENABLE US TO FILL ORDERS.

WE WANT THE BEST DEALERS....

To handle our complete line of goods, such as Geared and Pumping Wind Mills, Water Tanks, Feed Mills, Horse Powers, Pumps, Cylinders, Wood Saws, Shellers, etc. We will meet any competition when quality of goods is taken into consideration. Address,

CHALLENGE
Wind Mill & Feed Mill Co.

HALLADAY STANDARD PUMPING WIND MILL.

14 Sizes, 8 to 30 Feet Diameter.

Guaranteed the best regulating, safest in storms, most powerful and most durable Wind Mill made. Manufactured by us with improvements for 30 years.

The illustration below represents the Halladay Standard Wind Mill as used for pumping water on stock farms.

This mill is adapted for pumping water for all purposes, including irrigation and drainage.

Halladay Standard Wind Mill, with Sails Spread and at Work.

REASONS WHY FARMERS SHOULD PURCHASE
THE HALLADAY STANDARD WIND MILL.

It continues to be manufactured by the same company who have made it a specialty for over thirty years, and who have added, from time to time, many valuable improvements. It was the first reliable self-regulating Wind Mill ever made, and continues to lead the van. It has stood the test more than a quarter of a century in all the States of the Union, and is used in almost every country of the world, and on the islands of the sea, and has gained a reputation as broad as its use is extensive. It is the only Wind Mill generally adopted by the leading railroads of this and other countries. It is the cheapest Wind Mill on the market, when power, workmanship and material used are considered.

HALLADAY STANDARD PUMPING WIND MILL.

WE AIM TO MAKE THIS MILL THE BEST AND NOT THE CHEAPEST.

The illustration below represents the Halladay Standard Wind Mill as used for pumping water for ornamental purposes, such as sprinkling lawns, running fountains, etc.

By placing a Tank in the top of the Wind Mill Tower, the water may be conveyed to the barn, house or any point desired, and used for watering stock, washing carriages and windows, sprinkling lawns, running fountains, fire protection, etc.

Halladay Standard Wind Mill, with Sails Furled and at Rest.

The United States Wind Engine & Pump Co., who manufacture the Halladay Mill, have the largest Wind Mill Manufactory in the world. The enviable reputation that this Mill has earned, provokes the attacks of competing manufacturers who are offering machines the merits of which will not recommend them. All we ask is a fair investigation. We are willing to put our Mill upon its merits. Before credit is given to the erroneous statements made and published in circulars, by competing manufacturers, we ask an investigation of facts and motives.

We duly appreciate the patronage and support we have received in the past, and while we solicit a continuance of favors, we shall try to merit them. As we are the pioneers in the production of a Wind Mill adapted to the use of the farmer, we may be pardoned for suggesting that we have added millions to the value of this prairie country.

GAY & SON, OTTAWA, ILL. Manufacturers of **ROAD CARTS.** Best made. Positively no horse motion. Send for free Illustrated Catalogue of eight different Styles.

BATTLE CREEK ROAD CART.
EASIEST CART IN MARKET.
SINGLE, DOUBLE and SPEEDING.
SIMPLE, STRONG AND DURABLE.
Send for Description.
BATTLE CREEK ROAD CART CO., Battle Creek, Mich.

THE NOYES ROAD CART!

Manufactured by
THE NOYES CART COMPANY
KALAMAZOO, MICH.

Some of the Advantages of the Noyes Cart over all others: There is no bar for the rider to climb over in getting in or out. The horse can be hitched eighteen inches nearer than on any other Road Cart made. You can get in or out with perfect safety—no danger of being thrown into wheel—a great advantage in breaking colts.

The Easiest Cart for Road or Track!

Highly endorsed by leading horseman. All our carts are made of the very best materials. They sell well. Special inducements to dealers. Write for prices. Mention this paper. Address,

NOYES CART CO., - - Kalamazoo, Mich.

THE OTTAWA WAGON.

Finely Finished, Well Proportioned, Light Running, Durable, Superior Quality, Unexcelled. . . . . . .

Also Manufacturers of the Celebrated "Ottawa" Cylinder CORN SHELLERS, HORSE POWERS, etc.

Send for Catalogue.

Mention this paper.

KING & HAMILTON CO., - Ottawa, Illinois.

American Gear & Spring Co.

American Road Wagon—BUFFALO BODY.
Write for Catalogue and prices.
1445 Niagara St., BUFFALO, N. Y.

NORTH-WESTERN SLEIGH CO.,
MILWAUKEE, WISCONSIN, U. S. A.

Price, $20.00 F. O. B. Milwaukee.

YANKEE CART, NO. 6.

OUR Young Men's Buggy

CAN furnish either end spring or side bar, plain or biscuit style cushion and back. **We can prove this to be the most popular buggy on the market.** Our sales to date on the YOUNG MEN'S style are OVER Eight Thousand jobs. This buggy will hold two passengers,

And is Just What the Young Men Want.

It is built **LIGHT** and **STRONG**, with swelled body and seat, ¾ wheels, ¹³⁄₁₆ axles and light wood work. **WRITE FOR SPECIAL PRICES.** ☞ MENTION FARM IMPLEMENT NEWS.

SAYERS & SCOVILL.

Factory at Brighton Station,
Post Office Address, 51-61 Colerain Ave.,
Cincinnati, Ohio, U. S. A.

EVERY BUGGY WE TURN OUT is taken through our 100 DAY SYSTEM of Painting before Ironing. Read Our Big Point Circular. . . . .

BALL & WARD, Newark, Ohio, 82 Church St.

STICK ROAD WAGON.
We are making the best Wagon in the market. Special inducements to Dealers. Write for Prices. Mention this Paper.

CARTS, WAGONS, SLEDS and CARRIAGES. There was only one blacksmith among the first Jamestown Colonists and he died the next summer; so we assume that the pilgrims made do with the traditional poor man's wagon, a *sledge*. A passenger sled is a winter conveyance, usually of much lighter construction than the more utilitarian *sledge*. The most primitive farm sledge was called a *tumbril* and consisted of two ten-foot poles with crude runners carved on their bottom ends. Four upright poles and several cross beams held it all together and kept hay and cornstalks from sliding off the back when the front handles were elevated to be pulled, travois-style, by the farmer or his stout wife.

By the 1750s a *two-man sledge* had evolved, consisting of a wagon-style box mounted on four-inch wide runners. This much heavier sledge could haul anything from maple syrup barrels to a dead horse. Later, around 1780, came the dog sled-style *harvest sledge* with two split-pole runners and tall upright spindles on each side. *Bob sledges* were basic all-purpose farm sleds about forty-inches long, consisting of a pair of runners doweled together to form an open frame with no bed. A pair, hooked in tandem, could haul a 3,500 pound load over snow covered ground.

When you slide a heavy piece of furniture across the room on a rug, or give your child a ride on the flat end of a broom, you can understand the sledge's value as a wheel-less cart. The sledge could move anything from apples to animals, rocks, logs, etc., over rough land, through the forest, across ice and snow, by horse, oxen or human power -- often in places a cart or wagon could not traverse. Artist/historian, Eric Sloane, said that for every wagon on an Early American farm, there were from three to ten sledges.

The first *two-wheeled farm carts* used in the Colonies did not resemble the tall, iron-rimmed, spoked-wheel farm carts, left behind in England. Ox carts used by the French settlers of Illinois, prior to the War of 1812, were very primitive, being made entirely of wood without any iron fittings or nails. Some early American generic paintings portray low-slung carts with solid wooden wheels.

Four-wheeled farm wagons were made by professional wainwrights (wagon makers), wheelwrights (wheel makers), and blacksmiths (iron mongers). These specialist craftsmen engaged in trades that date back at least 3,000 years. Spoked wheels of light construction were being made in China, Egypt and Greece as early as 1500 B.C. The Celts were making bent-wood wheels in 500 B.C. that closely resemble those of the 20th century. The early Romans had carts, chariots and wagons, but the use of wheeled vehicles ended abruptly with the collapse of the empire and was not revived until the mid 16th century when an enamored Dutchman presented Queen Elizabeth with a fine heavy coach for her personal use. British nobility and other royal admirers soon followed suit and a new industry was born that eventually provided employment for thousands of English workers.

Obadiah Elliot invented the elliptic spring in 1804, and later another designer added leather shock absorbers. Within a short period the hard-riding cumbersome coach evolved into a graceful vehicle. However, its use in America was very limited. Roads were few and poorly maintained, and only the very wealthy could afford a carriage. In 1798, just one hundred and fifty carriages could be found in all of Boston.

Conestoga wagons were the "big-rig" trucks of 18th century America. Contemporary writers reported that the roads between some cities were constantly clogged with these giant freight carriers; as many as 3,000 wagons ran weekly between Pennsylvania farm towns and the market cities of Philadelphia and Baltimore.

Conestogas were drafted into service during the French and Indian War, and later George Washington used them to transport supplies to his troops during the American Revolution. After the war Conestoga wagons were used to carry manufactured goods across the Alleghenies to frontier settlements and return home loaded with western produce.

A six-horse Conestoga, with its 6-foot diameter rear wheels and iron rims up to 10-inches wide, could easily move six tons of cargo across a thousand miles of trackless prairie sod -- about a ton of cargo for each horse. These horses, usually from four to seven in number, were carefully matched in color and their heavy broad harnesses were of the best leather, trimmed with shiny brass plates. Even the bridles were brightly decorated with ribbons, rosettes and ivory rings, and each horse had a full set of musical bells fastened to a metal arch above the hames.

The wagons were brightly painted, usually blue with vermilion sideboards. The sixteen-foot long, boat-shaped box, with its narrow bed, was designed to keep heavy loads centered while traveling over steep mountain terrain.

These glittering rigs, from 40 to 60 feet long, with cracking whip and chiming bells, were indeed an impressive sight to behold. Drivers walked along beside the wagons, or rode the left wheel horse, controlling their intelligent animals with precision and ease.

Prairie Schooners that carried John Deere's plows out west were smaller, lighter versions of the Conestoga. These graceful ships of commerce, with their white canvas bonnets, were a striking picture as they moved over the hills in long winding trains. Their heyday ended about 1850, after which time railroads became the prime movers of freight.

The carriage and wagon trades were quite firmly established in the United States after the War of 1812. Most were small concerns, located within a fifty-mile radius of their markets. According to the 1870 census, there were at that time, 141,774 blacksmiths and 42,464 carriage and wagon makers plying their trades in this country. The U. S. Patent Office reported that 7,632 establishments were making carriages in the year 1900.

Carriage design was an art as well as an engineering feat. Wealthy patrons had custom-made carriages built to their individual specifications. A 1/12-scale plan was first drawn on paper, showing several views of the vehicle. After the customer signed off on this set of paper plans a precise full-scale shop drawing was executed on a huge blackboard. Pattern makers from various departments would obtain copies from the board by pressing damp paper or cloth against the chalk outline; from these patterns a wooden framework emerged that then moved on to a metalworking department where hinges, bolts, locks and handles were attached.

THE IRON CLAD, white ash framework was now ready for the body department where it received its floor, dashboard, door panels and seats. From there the body went to a paint room where ten or fifteen separate coats of lead-based paint and primer were sanded and hand-rubbed to perfection with pumice stone.

After a final color and varnish coat the body was rolled into a trimming room to be decked out in mohair, morocco, broadcloth, and velvet upholstery, carpet and trim. While the body was thus being prepared the wheels, axles, perches, steps, springs and shafts were being readied to receive it. After the installation of a hand-buffed leather top and a final polishing job, this work of art was ready to travel the muddy, pot-holed, manure-strewn roadways of 19th century America.

The wagon business entered the mass-production era sooner than the carriage-making trade. After the Civil War most farm wagons evolved into single, double, and triple box-styles, with spring-mounted seats. A triple box wagon was made up of three sections, 14, 12, and 10 inches deep. The second section was called the top box and the third was named the tip-top box. The more prolific, double box-style, consisted of the main box, supplied with an extra set of sideboards. A common single-box farm wagon measured a bit more than ten feet long and three feet wide, and held about two bushels of grain for every inch in depth. The bed could be removed and a hayrack installed, or the running gear (called a *truck*) could be used independently to haul logs, lumber, or agricultural machinery.

Prior to 1918, wagon tracks were 60 inches wide. But in the 1920's, when automobiles with their 56-inch wide tracks began to outnumber wagons by ten to one on the country's mostly unpaved roads, International Harvester Co. offered a 56 inch "auto-track" wagon that would roll faster in the established ruts of the horseless carriages.

Three or four inch thick planks of white ash, white oak, or yellow poplar were the basic lumber used in wagon and carriage construction. Wheel hubs were turned from sour-gum, white rock-elm or black locust; which was cured for five years before turning and inletting spokes made of hickory or oak. A first class Philadelphia body shop of the 1870's kept 80,000 board feet of lumber on hand in its drying rooms.

A well-made wagon would have all its wooden *wheel parts* boiled in linseed oil, and a coat of paint applied, before the iron tires were shrunk to a tight fit by heating and then plunging the assembled tire and wheel into a trough of cold water. After "ironing", two more coats of red lead paint were applied, then pin stripes, and finally a coat of varnish. The rest of the wagon received a similar three-coat paint job, usually ending up with a green box and red wheels.

Wagon manufacturing methods had become refined to the point that two companies were able to provide 550 wagons for the U. S. Government's Utah expedition of 1857 in only 5 weeks — a new wagon completed every 45 minutes!

The STRENGTH of the
STUDEBAKER

¶ The secret lies (for the most part) in the WHEELS and AXLES. That's where the *real* strain falls on a wagon. Those are the parts that must demonstrate success or failure. For no matter *how* good your wagon may be in *other* respects, it will not stand up to the work if the wheels and axles are *weak*, or the wheels improperly "set" or "gathered."

Wagon and carriage making were the 19th century equivalent of today's automotive industry in their importance to the overall economy. In fact, several automobile makers got their start in the carriage business. The Durant-Dort Carriage Company's founder, Will Durant, also founded General Motors Corporation, and the Studebaker Brothers applied their manufacturing skill to producing automobiles.

During the stampede of emigrants to California and Salt Lake, Utah, in the 1850's, the demand for wagons sometimes exceeded the supply. Wagon makers in many cities became rich beyond their wildest dreams. Factory workers made a dollar a day and put in ten to twelve hour shifts, six days a week. There were labor strikes, industrial accidents, lead poisoning and lung disease; plus taxes, tariffs, insurance and freight regulations. The "good old days" were more hectic than we tend to imagine. Competition was fierce, there were dealers, drummers, jobbers, trade journals, trade schools, trade shows, and international exhibitions.

Blacksmiths were the equivalent of today's auto mechanics. Wagon and carriage parts frequently failed and their iron tires had to be re-shrunk every few years. The horse-and-buggy era provided a good livelihood for many other businessmen, craftsmen, and laborers. There were livery stables, hay farmers, horse breeders, farriers and feed stores, as well as harness, horse collar, buggy whip and saddle makers. What a shock it must have been to all these tradesmen when automobiles took over American roadways at the close of World War One.

The average farmer of the prewar period felt fortunate to own his wagon and a team of workhorses; the purchase of a carriage was out of the question. Before the Civil War it took a year's salary to pay for a custom-made carriage, even economy models were priced well over one hundred dollars. However, by the early 1900's more than 100,000 pleasure rigs a year were rolling off the assembly lines and a basic buggy with a folding top could be ordered from Sears & Roebuck for $34.95 (about one month's wages).

Those folks not wishing to be saddled with the upkeep of a horse and buggy could rent an animal, or vehicle of their choice from the local livery stable.

At the turn of the century there were 7,632 carriage and wagon making factories in the United States, ranging from small buildings to fifteen-acre complexes, such as that of the Studebaker Brothers, located in South Bend, Indiana.

John Studebaker, the son of German immigrants, was born in 1799, near Philadelphia. After marrying a local girl, he learned the blacksmith's trade, built himself a Conestoga wagon, and headed west to Ohio. John was not successful financially, having to support a wife and six children by marginal farming and some forge work; but his eldest sons, Clement and Henry, formed the Studebaker Blacksmiths and Wagonmakers Company in 1852, with sixty-eight dollars in capital. By 1870 they had hired fourteen workers and were worth a cool ten thousand bucks. In the year 1872 they were "The Largest Vehicle Builders in the World", with an annual capacity of 75,000 coaches, carriages and farm wagons.

"ECLIPSE" FARM ENGINE.

HIGHEST CENTENNIAL AWARD and SPECIAL MENTION by the U. S. CENTENNIAL COMMISSION.
A MEDAL AND TWO DIPLOMAS.

INTERNATIONAL EXHIBITION, PHILADELPHIA, 1876. No. 235.

The United States Centennial Commission has examined the report of the Judges, and accepted the following reasons, and decreed an award in conformity therewith:

Philadelphia, January 31st, 1877.

REPORT ON AWARDS.

Product, *Portable Farm Engine.* ("*Eclipse.*")
Name and address of Exhibitor, *Frick & Company, Waynesboro, Pa.*
The undersigned, having examined the product herein described, respectfully recommends the same to the United States Centennial Commission for Award, for the following reasons, viz.:

This engine gives the best results of any that were tested, and may be regarded as a well-made, strong, and useful machine. The traveling wheels are large and powerful. The boiler is suspended on springs for traveling, which are let down when at work. The boiler is capacious. There is a powerful brake on the hind wheels, very useful for staying the engine when at work. The engine is carried on the top of the boiler, resting on a powerful bed-plate, which is hollowed out to form a receptacle for oil leakage. This can be detached from the brackets, and the engine converted into a fixed horizontal engine if required. The governor has three speeds, and the crank shaft is counterbalanced. The engine saddle has provision for varying expansion. The water heater is large, of the ordinary diaphragm form, and the pump with air chamber is well constructed. The cylinder has balanced slide valve. The safety valve works by a spring, which is a good arrangement, particularly when the roads are rough. Driving wheel on each side of crank shaft.

GEORGE E. WARING, Jr.,

Cincinnati Industrial Exhibition, 1874; Maryland State Fair, Baltimore, 1874; Virginia State Fair, Richmond, 1874; North Carolina State Fair, Raleigh, 1875; Delaware State Fair, Middletown, 1875; Pennsylvania State Fair, Lancaster, 1875; North Carolina State Fair, Raleigh, 1876; Georgia State Fair, Atlanta, 1877; Maryland State Fair, Westminster, 1877; also numerous County Fairs, and wherever else exhibited.

MANUFACTURED ONLY BY
FRICK & CO., WAYNESBORO, PA., U. S. A.

70 MILES WEST OF BALTIMORE, ON WESTERN MARYLAND R. R.
ALSO, MANUFACTURERS OF

STATIONARY ENGINES and BOILERS, CIRCULAR SAW MILLS, Etc,
For Particulars of the Above Send for Illustrated Catalogue.

Massive self-propelled steam engines, with rear drive wheels from two to ten feet wide, arrived on the scene in the 1890's. These farm locomotives were developed to replace the 32-horse teams that pulled the giant threshers, combined harvesters and gang plows on huge "Bonanza" farms comprising from 7,000 to 70,000 acres.

Especially suited for this scale of operation were the Great Plains states and the vast broad valleys of Washington, Oregon, Central California and North Dakota; where the land lay nearly flat as far as the eye could see.

According to Mitchell Wilson, author of *American Science and Invention,* these Bonanza farms yielded twenty bushels of wheat per acre. The average 7,000 acre spread was worth about $200,000 including its inventory of ten 4-horse plows, eight seed drills, six harrows, seven grain binders, and three steam-driven threshing machines. One such farm even had its own railroad that ran across the field.

"A standard feature of every bonanza farm was the grain elevator, which had first been introduced in 1842 so that the grain could be stored against market fluctuations. The average Red River Valley, North Dakota, farm had elevators with a capacity of over 100,000 bushels."

"Crews working at one end of the farm might not see crews working in other sections from one season's end to another. Fifty men were employed during the plowing season; forty of them were discharged when the season ended. The cost of growing and yielding an acre of grain, including labor and the latest mechanical equipment was $3.75."

Time saved by the new machines eventually drove grain prices to a level that would no longer support such far-flung operations. Sharp increases in the wages of farm laborers, and other overhead expenses eventually took their toll. The largest steam engines of the day consumed more than a cord of wood per day and about a ton of water an hour and required a full time crew of two men, two horses and a water boy. Coal was the most favored fuel and ran up to $14 a ton. Farmers were waiting for smaller, lighter, cheaper-fueled engines and within two decades gasoline-powered tractors had made traction steam engines obsolete.

By the turn of the century most of the ultra-large bonanza-sized holdings were profitably converted back to family-sized farms that relied on horses and mules for power.

During the Golden Age of agriculture, from 1909 to 1914, steam engines were providing power equal to 7 million horses and mules. The farmers of a generation earlier could not have imagined the possibility of raising wheat on this grand, western scale. It was not a question of square acres, but of square miles; not of bushels, but of train carloads.

In 1901 the wheat crop of the United States was 756,269,573 bushels, an amount equal to one fourth of the entire world's production. Prices were subject to wide fluctuations. During the three years prior to 1903 they ranged from 61 to 95 cents a bushel. Grain exports for the season of 1901 amounted to over sixteen million dollars.

For a more complete history of steam engines used on the farm, please see pages 113 to 118.

Case 30-60 Tractor

Case 20-40 Tractor

A Size for Every Farm —

30 H. P.

40 H P

50 H. P.

Case Builds Eight Different Sizes of Case Steam Engines — 30 H.P., 40 H.P., 50 H.P.
Ever Entered Case Has Always Taken Highest Honors —

J. I. CASE offered these tractors in their 1918 catalog.

Case 12-25 Tractor   Case 10-20 Tractor   Case 9-18 Tractor

very One an Economical Kerosene Tractor

65 H. P.   80 H. P.

) H.P., 65 H.P., 75 H.P., 80 H.P., and 110 H.P.—In Every Steam Contest
nexcelled for Simplicity, Durability, Economy and Accessibilty

J.I. Case 9 and 10 hp. Kerosene Tractors of 1918.

Case 9-18 Kerosene Tractor
A Practical and Popular Size—For Both Large and Small Farms

Case 10-20 Kerosene Tractor
An Excellent 3-Plow Tractor with Plenty of Reserve Power

J.I. Case 9 and 10 hp. Kerosene Tractors of 1918.

Case 12-25 Kerosene Tractor

A General Purpose Tractor—It Pulls Four Plows

Case 20-40 Kerosene Tractor—Noted for Its Durability

It Pulls a 6-Bottom Plow

MINNEAPOLIS 25 H.P., 4-cylinder (vertical), gasoline tractor of 1914.

"The Minneapolis 25 Kerosene Tractor" Plowing on the Farm of "D. Henderson" Three miles west of Boissevain, Man. Fall 1913.

MINNEAPOLIS 40 hp., 4-cylinder (horizontal), gasoline tractor of 1914.

Minneapolis Farm Motor Threshing in the Red River Valley.

Illust. 1—Model G, International Motor Truck—capacity 2 tons. Equipped with regular express body and cab top.

Illust. 1—Weber Central States farm wagon. These wagons are built in types to suit all sections of the country.

International Harvester Company, 1920.

Illust. 1—The new International roller-bearing, tight-bottom manure spreader.

Illust. 1—McCormick Special huskers and shredders are made in 6, 8 and 10-roll sizes. They have small husking rolls

Illust. 2A—McCormick No. 6 mowers are used in hay-growing countries all over the world.

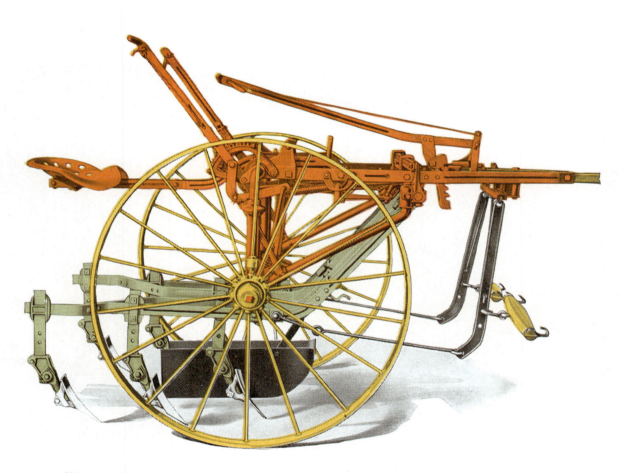

Illust. 1—International No. 4 pivot axle cultivator. The most popular riding cultivator made. Equipped with six shovels, pin break, round shank.

Illust. 1—P & O Little Genius No. 5, three-furrow power-lift tractor plow.

Illust. 1—Hoosier Easy Pull fertilizer grain drill with 12 open delivery, single disk, furrow openers.

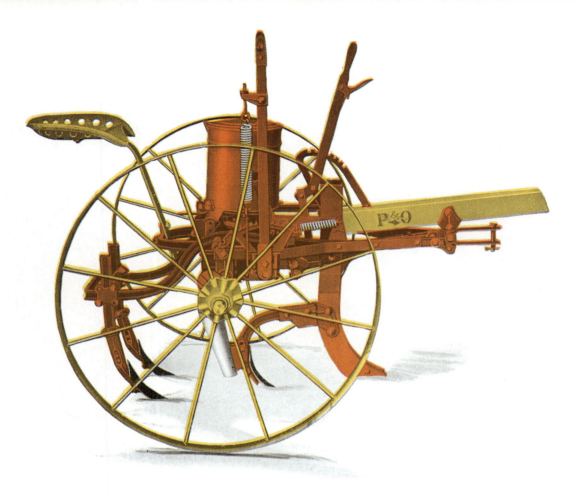

Illust. 44—P & O No. 1 Tip-Top cotton and corn planter.

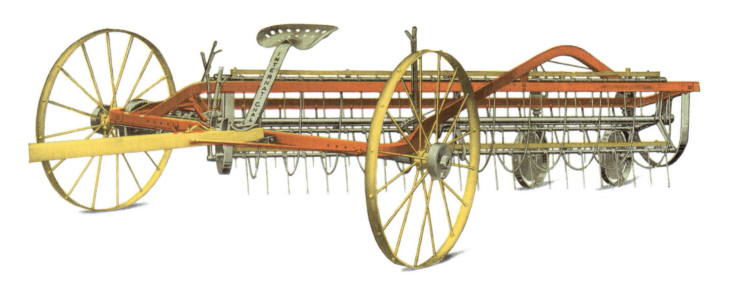

Illust. 1—International combined side rake and tedder, with extra caster wheel.

Illust. 1—Deering binders will harvest grain when conditions are good, and have adjustments to meet many conditions when grain is difficult to harvest. Made in three sizes, with 6 7 and 8-foot cuts.

Illust. 1—McCormick Improved grain binders are backed by nearly a century of manufacturing experience. They are made in sizes to meet the needs of every grain grower—6, 7 and 8-foot cuts.

Illust. 1—International power press equipped with 6 H. P. International kerosene engine.

Illust. 1—International 8-foot disk harrow, equipped with forecarriage.

Case 20x36 Lightweight Steel Threshing Machine

Also Built in Sizes of 20x28 and 26x46—Real Grain Savers

Case 32x54 Steel Threshing Machine—Six Other Sizes

It Represents Three-Quarters of a Century of Threshing Machine Building Experience

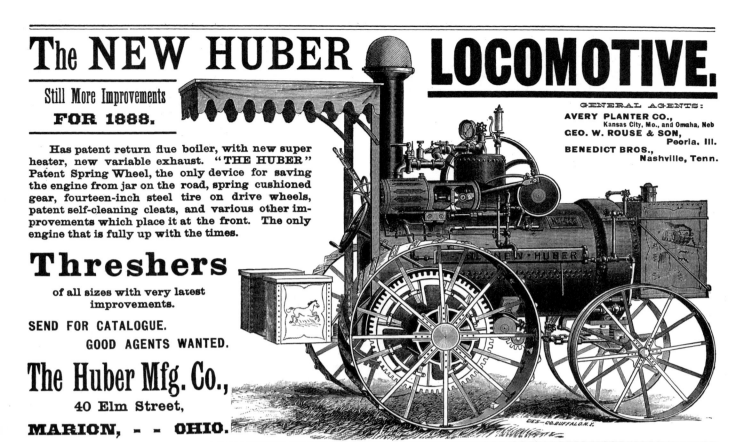

STEAM ENGINES. *Self-propelled traction engines,* (steam tractors) evolved from two sources during the 1868 to 1875 period: Farmers who had their portable *agricultural engines* converted to primitive tractors, and the adaptation of *road locomotives* to farm work.

Road locomotives were so named because they could propel and guide themselves on common roads without rails. They were sold in different localities for variety of chores, including road building, rock hauling, lumber mill work, and the transport of cotton and sugar cane to railway stations.

The idea of using road locomotives for farm work originated in England where Wm. Fowler & Company, and a few other firms, had been building these heavy machines for wealthy landowners in Great Britain's many colonies. These converted locomotives were anchored across from each other in a field where they drew a gang-plow back and forth between them. The system employed 800 feet of steel cable, strung between a set of oversized pulleys and anchor blocks. This steam-driven plow operation ran faster than a man could walk and cultivated between fifty and sixty acres a day, as compared to two acres per day plowed by a team of horses.

But most American farmers did not have the $8,000 price tag of a road locomotive and they balked at the idea of using such heavy equipment in their muddy and hilly fields.

A much lighter machine was needed that could be hitched directly to a plow and operated by just one or two men. Ultimately, steam tractors proved to be practical only on large, dry, flat acreage. Even then, their voracious consumption of both fuel and water made them obsolete when gasoline engines were perfected in the early 1900's.

An 1892 article in *Farm Implement News* **noted that the** largest steam engines of the day used about a ton of water per hour and even an average sized engine required a full-time water boy and a team of horses to keep it supplied. On top of that was the cost of fuel; coal prices ran from $7 to $14 a ton. The author stated that gas engines should be developed soon.

A forgotten Ohio farmer was the first American to patent the idea of a horse-steered, self-propelled, steam traction engine. In the summer of 1869, an Ohio farmer hired the C. & G. Cooper Co., of Mount Vernon, to convert his agricultural steam engine into a wheel-driven tractor that could be steered by two horses attached to its tongue.

Cooper and Company saw the potential for profit right away, and offered to do the entire conversion job for nothing...if the farmer would secure a patent and sign it over to Cooper as payment for services rendered.

In 1873, John Yingling, of Seven Mile, Ohio, hired Owens, Lane & Duyer Co. to convert his ten-horsepower steam engine into a chain-driven tractor. They first used a pair of oxen to steer the heavy contraption, but it began rolling so fast down a hill that it ran over the team. After the accident it was converted to self-steering by means of a hand-cranked, chain-wound, drum.

To make a long story short, C. & C. Cooper continued to buy up all the tractor-related patents they could lay their hands on, and by the late 1870's they were marketing their engines all over the country. They were also receiving royalty fees from other manufacturers, including Russell & Company, Owens, Lane & Duyer, E. M. Birdsall Company, and the Wood Brothers.

C. Aultman & Co's Monitor Engine.

Monitor Traction Engine.

THE C. AULTMAN & COMPANY was one of three agricultural machinery manufacturers formed by the stepbrothers, Cornelius Aultman and Lewis Miller, between 1849 and 1865. Cornelius was apprenticed to a wheelwright as a teenager, and became quite an accomplished machinist before the age of eighteen. One of his first jobs at Ephraim Ball's Greentown machine shop was to duplicate Obed Hussey's reaper for the local market. A neighboring farmer, named Michael Dillman, bought one of the copies made by Ball and Aultman, and was so impressed with its performance that he decided to move to Illinois and start mass-producing the stolen design for a much wider market. He hired Aultman and his stepbrother, Miller, and opened a shop in Plainfield, Illinois, in 1849. Meanwhile, Obed Hussey heard of the illicit enterprise and rushed from Baltimore to confront the copycats. The case was settled out of court for a royalty payment of $15 per reaper and Hussey hit the road in pursuit of other infringers – a job that kept him busy for many years. To make a long story short, Aultman and Miller went on to build more of Hussey's machines as well as several mower designs for the 1853 harvest season. In 1855 they added Pitts' threshers to the line and in 1856, invented the famous Buckeye mower, which was also licensed to other builders. By 1863 the brothers had bought out a partner and opened a second factory that produced combined sales of 8,000 Buckeye mowers and 500 Sweepstakes threshers. In 1865, Aultman moved to Mansfield, Ohio, and started a third enterprise; building threshers, horse powers, clover hullers, saw mills and steam engines. Along the way several relatives and other investors took part in the three factories. Both brothers had become independently wealthy by the start of the Civil War. (This page and the next are from an 1889 catalog.)

THE PHŒNIX ENGINE.

Phœnix—Without Wood Casing.

OUR STRAW-BURNING ENGINE.

Only a few years have elapsed since the cry for straw-burning engines became so pressing that it had to be heeded. Fortunes were sacrificed in experiments. Five years ago C. Aultman & Co. touched rock bottom with the Phœnix. It has been an entire success from the start. The correctness of its principles has been confirmed, and its reputation for superior excellence strikingly established by the conduct of our competitors in abandoning their faulty contrivances, and falling into line with imitations bearing their own names.

One feature of our trade in Phœnix engines has both pleased and surprised us. The first year this engine was included in our catalogue several machines were selected and bought to be used as wood burners. Since that time a considerable number have been sold to be used with wood or coal. To say that they have, in every instance, given entire satisfaction, falls short of the truth. The performance of these machines, whether fired with wood or coal, has been mentioned only in terms of the highest enthusiasm.

Buffalo Pitts
Steam Traction Engines for 1891

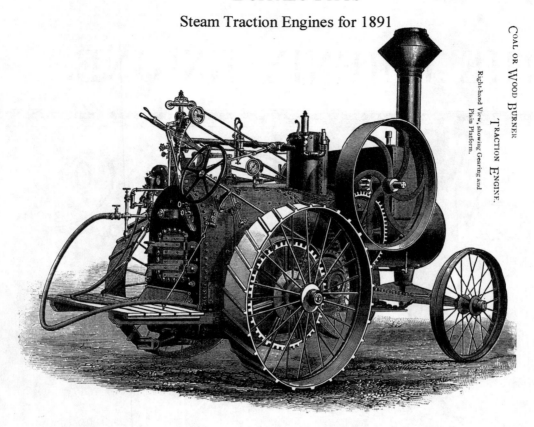

Coal or Wood Burner Traction Engine.
Right-hand View, showing Gearing and Plain Platform.

Coal or Wood Burner Traction Engine.
Right-hand View, showing Gearing and Plain Platform.

BUFFALO PITTS COMPANY was established in 1851 by the brothers Hiram and John Pitts, who had patented the first practical threshing machine in America in 1837. The brothers began manufacturing steam engines in 1881, and by 1891 were offering a line of four different coal, straw, and wood burning traction engines from six to twenty horsepower, (see page 70 for more on the Pitts).

In 1882 Russell & Co. brought out one of the first self-steered traction engines, but doubting farmers had to be convinced that horses were no longer needed for steering. Russell's creative advertising team was up to the challenge:

"We recommend our self-steering engine to all who wish to be independent of horses. We do not expect them to take the place of horse steered engines, but we do predict they will be found useful in great wheat growing districts of the Northwest. Until the rights of the traction engine on the road are more thoroughly established, we do not advise their use in the more thickly settled portions of the country."

By the 1890's self-propelled steam engines were widely used on the Great Plains and on large farms in the Far West; many of them were capable of plowing up to fifty acres a day.

Steam engine boilers were powder kegs, just waiting for a careless operator to make the wrong move. The following story appeared in the April 2nd, 1893 edition of the Chariton, Louisiana, Herald: "On last Saturday afternoon the boiler of Henry McKinnis' sawmill exploded with terrific force, killing the proprietor and his three sons, aged 19, 12, and 10, who were working with him. The father was found 75 yards from the mill gasping in the last throes of death. The eldest son, John, was found some ten yards farther away, conscious, but in a dying condition. The 10 year-old boy, Fred, was found dead, near a pile of lumber in the yard, against which he had been thrown, with his skull mashed in. The body of Henry, Jr., aged 12, who was acting as engineer and fireman, was literally blown to atoms, pieces of which were found 300 yards away."

"The boiler was an old one and in leaking condition. Just before the explosion, a neighbor observed that the water was low in the boiler and that the pump was not throwing as it should. The 12 year-old engineer was trying to get more water going through, but apparently the boiler became dry and hot. One of the sons was seen walking back to the boiler and starting up the water pump, and just as he was walking away the explosion occurred. The noise was heard for miles away and farmers who rushed to the scene found nothing but shattered trees and mangled bodies. It is supposed that the explosion happened (as usual) when cold water was pumped into the hot, empty boiler."

Another sad story from the steam threshing era: "Lowell Klinefelter, 36 years old, was fatally injured Tuesday when the threshing machine into which he had just tossed a bundle of grain, grabbed his pitchfork and hurled it back at him, thrusting the handle through his skull."

Operating a steam engine safely required close attention to numerous details. The water level had to be carefully monitored at all times. If the water was not high enough in the boiler the resulting overheating could warp the metal, crack the joints, or blow up the boiler. When a serious leak or an abnormally low level of water was detected, the operator had to immediately cool down the boiler by covering the fire with ashes or dirt, or try to smother it with fresh coal. Any stirring or attempts to draw the fire could result in even more heat. The normal procedure was to leave an engine running during any attempt to douse the fire.

Since a steam engine is very hot when operating, it requires a lot of oil on every moving part. Most bearings needed several drops a minute and the main cylinder required an extremely heavy black oil to withstand its higher temperature. Starting up and closing down a steam engine required opening and closing a number of different valves, each in its proper sequence; nothing was done in haste.

Handling a traction engine on the road was no easy matter. This was especially true in hilly country. The gauge glass and water cocks had to be carefully watched and the steam pressure maintained near the blow-off point. "Upon approaching a hill judgment should be exercised in regard to the fire and amount of water. As much water should be carried as possible without priming. There should be sufficient fire when starting up a hill to carry the engine to the top. Also there is the danger of reducing the steam pressure so that a stop will have to be made to raise it. When the summit of the hill has been reached, the fire can be started up, more water put in the boiler, and the engine allowed to travel faster. As much or even more care must be exercised in descending a hill. The engine should be taken from top to bottom without a stop. If a stop has to be made you must turn the engine to a level spot. Every engineer knows the danger of having the front end of the fire box boiler lower than the rest of the engine. If the engine runs too fast in going down a hill, the reverse should be thrown. If it still travels too fast while in reverse, open the throttle and let in a little steam." (Words of wisdom from an old operating manual.)

The best way to get out of a mud hole was not to get into it. A wise engineer would go a long distance out of his way to circumvent a mud hole. The same methods were employed to get a steam tractor out of a mud hole that are used to free an automobile today. Lots of straw, brush, wood or fence posts, were thrust under the drive wheels; anything to stop them from spinning, and help the tires get a grip. If that failed a team of horses were hired to pull it out.

Heavy steam engines on their way to threshing jobs frequently broke through bridges and fell 20 or 30 feet into a swollen creek. If the crew was not drowned they were often maimed or killed by falling timbers, or crushed by the engine and its attached separator.

If an engine broke partially through a bridge, it might be saved by winding a rope around the power-belt wheel several times, and hitching a team to the rope. As the rope gradually unwound it moved the engine by a winding action upon the transmission gears, which were attached to the wheels. If a steam engine had fallen into a creek, it could sometimes be salvaged by a team of horses hitched to a geared-down stump-pulling machine.

Steam engines used in public places had to be inspected by a safety engineer, and all bridges were supposed to be posted with their maximum load capacity. The only insurance an operator could carry was a load of sturdy timber to re-plank any bridges that looked suspect.

All Case Steam Engines will develop at least 10 per cent more B. H. P. than rated.

Simple Cylinder—12-Inch Bore x 12-Inch Stroke

THE 110 H. P. size is the largest of the Case steam tractor family. It is designed for big farms and ranches where 10 or 12 bottom gang plows are used and where team operations are out of the question.

On these big acreages time is of vital importance. The work must be done quickly and of course, economically. There must be no delays for delays are expensive.

For this strenuous service the 110 horsepower engine is filling these requirements with that exactness that is making thousands of friends for Case throughout the world.

Of course, when it comes to belt operations, any agricultural machine, no matter what its size, can be operated by this "giant" power plant.

110 H. P. Steam Tractor Regularly Equipped

Pictured above is one of eight steam engines featured in the J. I Case Threshing Machine Company's huge 1918 catalog. Steam plowing and threshing continued in some areas until the 1930's, when gasoline tractors became widely affordable.

JEROME CASE built his first portable steam-driven agricultural engine in 1869, to power one of his company's already successful threshing machines. With its four-foot diameter flywheel it resembled a small locomotive mounted on wagon wheels and burned coal or wood (a cord a day). In a pinch, farmers could resort to brush or barley straw for fuel. Only 75 of the $1,000.00 engines were sold in 1876, but demand slowly increased and 3,163 units ranging from 6 to 50 hp were sold over the next ten years. By 1886, Case was the world's largest manufacturer of steam engines and sales of its popular "Eclipse" thresher propelled it to number one position in the worldwide market.

In its 1918 catalog, the J. I. Case Threshing Machine Co. featured five kerosene-burning tractors of 18 to 30 hp in addition to its established line of eight steam traction engines, ranging from 30 to 110 hp.

After several years of experementing, the Caterpillar's steam power plant was replaced with a gas engine in 1908. The awkward looking front wheel aided little in steering, but was needed to support the engine and front end.

BENJAMIN HOLT and his brother Charles traveled to California in 1880 from Concord, New Hampshire, where they had worked in their father's wholesale hardwood business. In 1883 the brothers established The Stockton Wheel Company to supply oversized implement wheels and agricultural machinery to farmers in California's vast Central Valley.

Out of that shop came *The Holt Bros. Improved Link Belt Combined Harvester*, which, according to an early 1900s magazine ad: "Cuts and threshes grain in one stroke".

The brothers also produced or distributed "Giant steam driven tractors to plow and harvest, plus multi-ton freight wagons, land levelers, harrows and sage brush plows — all on a bigger, scale than eastern farmers have ever known."

In the fertile delta soil the monster steam tractors sank down deep, no matter how wide their wheels; even with wheels up to sixteen feet wide on both sides of the machine.

In 1904, after years of experimenting with various wheel designs, Benjamin Holt substituted a continuous chain-belt of iron-linked wooden blocks, for the huge rear wheels on his steam traction engine.

The resulting crawler was named the Caterpillar. Although other track-laying, traction engine designs preceded it, Holt's invention was recognized as the first flexible self-laying track that could both support and propel a tractor under all conditions. It could turn in a shorter radius than other crawlers because each track had its own clutch and brake.

The Caterpillar's one-foot-wide track had more pulling power than an eight-foot-wide conventional wheel. It was just the ticket for drawing heavy implements such as threshing machines across the saturated Sacramento-San Joaquin delta soil, or up steep hillsides; and it could not be stopped by ice or heavy snow.

In 1908 Holt replaced the Caterpillar's bulky steam power plant with a modern gasoline engine (see photo above), and by 1910 the Holt Brothers were well established in the tractor business, with factories in Stockton, California, and Peoria, Illinois.

The caption on the back of the above photograph reads: Caterpillar tractor with a heavy steel casting for the rear drive sprocket. A pinion drive, working on teeth inside of this sprocket, makes possible its much lower center of gravity.

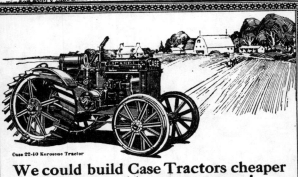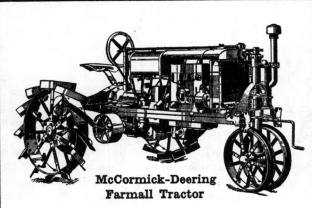

GASOLINE FARM TRACTORS evolved from steam-powered traction engines around the turn of the century. The Charter Gas Engine Co. of Chicago got a head start on competitors by simply mounting one of their hit-and-miss, single-cylinder engines on an existing Rumely steam engine chassis in 1889. They sold six of the giant rigs, with fully exposed gears, to large-scale farm operations in South Dakota. The Huber Mfg. Co., a steam engine-maker, mated a Van Duzen gas engine with their well-known steam traction-engine transmission in 1898, and sold thirty of the hastily conceived tractors in twelve months.

The J. I. Case Company began experimenting with gasoline engines in the 1890's and sold a few two-cylinder, gasoline-powered rigs in 1895. But it was not until 1912 that their first commercially successful model was introduced.

The 25,000-pound monster developed 30 draw-bar horsepower and 60 belt hp. But the market for machines of this size was limited and the next year Case came out with a smaller version, weighing only 13,000 pounds. Six kerosene burning models were introduced between 1913 and 1920, each being a smaller, lighter version of its predecessor. The 4-cylinder Case Crossmotor weighed a mere 5,700 pounds and out sold all other tractors in its class.

John Froelich, the owner of an Iowa-based agricultural machinery rental business, entered the fray in 1892, after observing some gasoline-converted steam tractors in South Dakota during a business trip. Froelich mounted a Van Duzen single-cylinder, 20-hp engine, with a bore and stroke of 14 inches, on a Robinson chassis and shipped it to South Dakota. In its first season the Froelich conversion ran for fifty days and threshed 72,000 bushels of wheat without a breakdown. This apparent success was not duplicated again by Froelich, or his partners, in the next two decades.

The Waterloo Gasoline Tractor Engine Co., of Waterloo, Iowa, teamed up with Froelich to produce several other experimental models, but farmers were unhappy with them and it was not until 1913 that the company achieved commercial production of the forerunner of the famous "Waterloo Boy".

The kerosene burning, two-cylinder, "Waterloo Boy One Man Tractor", developed twelve horsepower at its rear wheels and twenty-five horsepower from a belt pulley. Its $850 price tag and low cost of operation made the tractor a big hit with small farmers; more than 30,000 Waterloo Boys were sold between 1914 and 1924.

An Illinois farmer's son recounted his experience with this dependable machine in an issue of *Iron Man Album*: "About 1914 dad bought a 12-25 Waterloo Boy and a three bottom self-lift plow. It was a good little tractor in its day. In addition to plowing our forty acres we used it for all kinds of jobs around the farm. We also ran a sawmill and ground corn for several hundred head of cattle with that old Waterloo, which we sold in 1920."

Deere & Company paid over two and a half million dollars for The Waterloo Gasoline Tractor Engine Company, in 1918, in order to gain quick entry into the booming tractor market. However, on average, the change from horses and mules to tractors was actually a very gradual process. In 1910 there were less than 1,000 tractors competing with 19,420,000 horses and mules on American farms. By 1920 the number of farm tractors had increased to 246,000 vs. 22,386,000 draft animals. By 1935 the horse and mule population had declined to 15,471,000 while tractor numbers had grown to a million.

Some economists contend that the sharp reduction in the horse and mule population from 1920 to 1935 had released forty million acres for purposes other than growing feed for draft animals, and eventually contributed to lower farm prices because of increased grain production.

An agricultural textbook author stated in 1943: "Thirty years ago tractors and motor trucks were comparatively rare on farms, but now we have a tractor to about every 3.8 farms and a truck to every 6.5 farms. During the depression years of the early 1930's replacements of draft animals by tractors were at a slower rate than prior to, or following that time."

The fact of the matter was that only a few farmers had much experience with gasoline engines, or enough ready cash to purchase a tractor. Their horses were, on average, only about ten years old and could be maintained with leftover hay and grain, a few bushels of oats, and whatever else they could scrounge from pasture land.

Many farmers had a strong emotional bond with their four-footed friends who remained the chief motive power on most farms until after World War II. In several states, especially those with hilly terrain or low-income farms, horses and mules continued in use through the 1950's

The number of horses in the United States was about seven million in 1867 and rose to a peak of twenty-one million head on farms in 1913. As late as the 1940's about 72 percent of our six million farmers still used horses and mules. The 1,610,000 tractors then working the land numerically trailed the remaining eleven million draft animals. The average farm covered 174 acres and common wisdom of the day was that you needed to have at least seventy-five acres under cultivation before tractor power became much of an advantage. The suggested ratio of replacement was one tractor for every five or six horses.

In his 1917 book, *Equipment for the Farm*, Professor of Agriculture at Ohio State University, Harry C. Ramsower, wrote the following closing paragraphs: "The farm tractor situation is now in a very unsettled state. With so many new designs and sizes on the market and with all the extra features offered it is almost impossible to give any general advice to a prospective purchaser. The prevailing opinion seems to favor a high-speed, multi-cylinder, light-weight tractor, weighing from two to three tons, as the choice for a small farmer tilling up to 400 acres of land."

"As yet the tractor is not used extensively for any operation other than plowing and harrowing. A tractor can replace a certain number of horses, or it may provide extra power for the busy season. But it will seldom be found possible to dispense with all of the horses on an average farm because their combined power must be available during the peak seasons of cultivation, haying and harvesting. Since some of these operations are now being performed by tractors, it is possible that the gasoline-powered tractor will someday find a permanent place in agriculture."

By the time the professor's book was published, gasoline-powered tractors had already gained a permanent

foothold in the countryside. **Between 1910 and 1925 an unbelievable 250 manufactures were producing** one, two, and four-cylinder models, of 12 to 60 horsepower, weighing from one to ten tons.

Among them were colorful brand names such as: Advance-Rumely, Albaugh-Dover, Allis-Chalmers, Aultman Taylor, Avery-Horizontal, Bates Steel Mule, Bean Track-Pull, Bethlehem, Beaver, Big Bull, Caterpillar, Case-Crossmotor, Centipede, Cletrac, Common Sense, Creeping Grip, Dixie Ace, Eagle, Emerson-Brantingham, Farmall, Fordson, Frick, Grain Belt, Happy Farmer, Hart-Parr, Huber, John Deere, Little Giant, McCormick-Deering, Minneapolis Moline, Oliver, Sandusky, Steel Mule, Titan, Twin City, Uncle Sam, and The Waterloo Boy.

With so many brands on the market there were bound to be a few "lemons" purchased, and in those days there were no consumer advocates or money-back guarantees. However, in 1920 a Nebraska legislator, named Wilmot Crozier, bought a bogus "Ford" tractor (unrelated to the Ford Motor Co.) that lasted less than a week before it broke down. After plowing for a month or so with his new trouble-free Rumely Oil Pull, the senator decided it was high time somebody let the public know which makes were worth the money. The upshot was a grueling state tractor test with strictly enforced trail runs that were conducted under widely varying conditions. The results were made public in the spring of the same year, and one by one the lemon-makers were driven out of business.

In the 1920's you could buy a gas-powered garden cultivator for $185, or you could spend $6,250 for a deluxe "Yankee Ball Tread" tractor. There were about forty makes of *two-plow tractors* on the market, selling in the $800 to $1,595 price range. *Three-plow tractors* were the manufacturer's favorite size and they made 65 varieties of them, priced from $1,125 to $5,800. Twenty factories were turning out *four-plow tractors* in the $1,750 to $4,000 range. Huge tractors that could pull from *five to twelve plows* ran $5,000 to $10,000 each.

Henry Ford built an experimental tractor in 1906, and in 1910 started selling an "Auto-Pull" conversion kit that: "Turns your Ford into a powerful farm tractor". Several other firms made car-to-tractor conversion kits during the teens and for as little as $195 you could turn your Model-T loose in the field with a plow behind it. Another accessory was available that ran a belt from the fliver's rear wheel to power pumps, churns, feed mills, saws, washing machines and generators.

Near the end of WWI, Henry Ford decided that it was time to manufacture a real tractor, and set one of his best designers to work on the project. Eugene Farkas originated the idea of making the engine, transmission and rear axle housings out of extra thick castings, strong enough to bolt together into a solid unit that would eliminate the traditional frame and chassis design. The resulting "Fordson" tractor went into production in 1917, and seven thousand units were shipped to tractor-starved Britain within six months. The Fordson-F series was introduced in 1918, and the 2,700 pound, $785 machine, floored the competition with its low price and wide distribution. Nearly every Ford automobile dealership featured a shiny new tractor on its showroom floor.

In 1921 Mr. Ford decided to take an even bigger slice of the tractor business pie by engaging his arch rival, the International Harvester Company, in a price war that eventually knocked the Waterloo Boy and many other brands out of the game. Ford lowered his price from $785 to $395. And by 1930, only three dozen of the 180 firms that were making tractors in 1920 still remained in business.

International Harvester's 1924 introduction of the-"Farmall" tractor put a huge dent in Henry Ford's business. IHC's nimble 13 horsepower Farmall was the first low-priced tractor built especially for the cultivation of row crops. Its high-wheel, tricycle-design straddled the crop and each of its 40-inch tall rear wheels had its own brake, a design feature that allowed for very sharp turns in small fields. Truck farmers and dairymen from coast to coast lined up to buy them and 135,000 Farmalls were sold before 1932

Waterloo Boy
Three-Plow Tractor

12 H. P. at the Drawbar
25 H. P. on the Belt

Now $675.00
COMPLETELY EQUIPPED
F. O. B. WATERLOO, IOWA

NO EXTRAS TO BUY

Equipment includes Fenders, Wheel Lugs, Governor, Platform, Friction Clutch Pulley, Lever Controlled Adjustable Hitch. The Waterloo Boy comes to you complete for drawbar and belt work.

Years of Economical Performance Back of the Waterloo Boy Tractor

What W. E. Sorrells Says About His Waterloo Boy

Moore, Texas, October 5, 1921
John Deere Plow Company,
San Antonio, Texas.

Gentlemen:
Just a line to say a word or two for the Waterloo Boy Tractor I bought of you.

We have moved over roads that were next to impassable with it, have pulled heavy loads and climbed one hill with a full load that other wheel tractors go around.

Have had no trouble at all with starting it and can't say enough for it on belt; also, for burning kerosene it has no equal. I am doing heavy work, with 15 or 16 gallons of 8-cent oil instead of 22 cent gas. I feel this is quite an item saved when it is repeated every ten hours I run the tractor.

I worked with tractors of different makes of much higher cost price=than the Waterloo Boy and do not hesitate to say it has all I have ever seen beat from any standpoint.

I will be glad to answer any letter of inquiry that might be sent me.

Your friend,
(signed) W. E. Sorrells

Today on thousands of farms Waterloo Boy Tractors five, six and seven years old, and older, are still giving users economical and profitable service. There is a reason. The Waterloo Boy is designed for farm work—for the heavy-duty, continuous, day in and day out service that the farm tractor is called upon to perform.

Sturdy construction so necessary for long life and continuous duty, is built into the Waterloo Boy, together with simplicity that makes it easy to understand and easy to operate—a mechanic is not a necessity on a Waterloo Boy equipped farm.

Fuel costs run exceedingly low. Specially designed for burning kerosene, the simple two-cylinder motor utilizes this low priced fuel with real economy. In competitive fuel tests time after time the Waterloo Boy has led the field. Its fuel economy records are not found only in competition; on thousands of farms under actual everyday farm service, records for low fuel consumption have been made and are being made daily.

One big reason for the success of the Waterloo Boy on farms is its simple, low-speed, two-cylinder, heavy-duty motor. Having a large bore and long stroke, this two-cylinder motor develops 12 H. P. at the drawbar and 25 H. P. on the belt at the low speed of 750 R. P. M.

There are fewer motor parts on the Waterloo Boy to wear out, and because of its low speed the life of these parts is greatly increased.

Hyatt roller bearings—bearings that result in longer life, less friction and less draft—are used in all important bearing points.

The upkeep costs on the Waterloo Boy are unusually low. Mr. Eli Thompson, Jr. of Hicksville, Ohio, R. 6, is only one among many that will verify this statement. In a letter dated April 28, 1922, he says—"I have a Waterloo Boy Tractor. It has been out about five years and I haven't spent a cent for repairs and have used it for all kinds of work."

Repairs can be made on the Waterloo Boy more easily because of its unusual accessibility and because of its simplicity.

Remember this, the Waterloo Boy is a three-plow tractor. It is guaranteed to develop 12 H. P. at the drawbar. At the belt it developes 25 H. P. for operating your belt machinery—an important point to bear in mind.

At the new low price of $675.00 f.o.b., a figure far below the pre-war price, the Waterloo Boy will make an unusually profitable investment. Where can you get any other heavy-duty farm tractor at so low a price with the same power, with the many advantages, and with a record for economical performance that extends back over a period of years.

Act today. See your John Deere dealer or write direct to us for information in detail which includes our four-color booklet that tells all about this remarkable tractor.

JOHN DEERE PLOW COMPANY
MOLINE, ILLINOIS

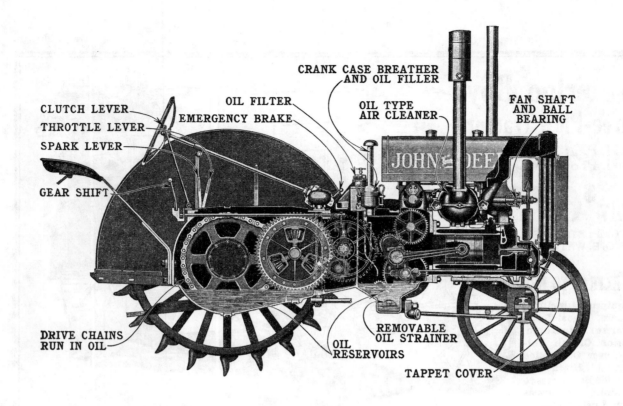

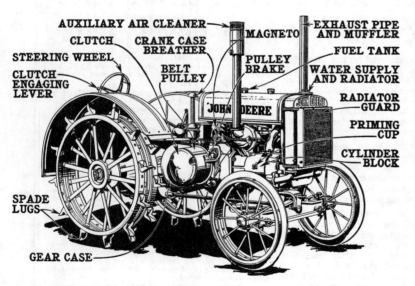

Figure 179, (facing page, top left) is a circa 1928 - 1935 two-cylinder General Purpose Model GP tractor. The drawbar horsepower was 10 and the belt pulley developed 20 hp. In the 1930's a larger version of the kerosene-burning engine increased its drawbar hp to 16. The 112-inch-long tractor weighed 3,600 pounds and sold for $1,200 in 1928. Over 30,000 units were produced.

Figure 180, (facing page, top right) is a tricycle-style model GPWT wide tread row-crop version introduced in 1929 to compete with International Harvester's hot-selling Farmall. The John Deere GPWT had a 76 in. wide rear tread which allowed it to straddle two crop rows. It sold new for $825.

The John Deere Model D (above) was produced from 1923 to 1953. The price of a new two-cylinder model in 1924 was $1,000. The kerosene engine developed 22.5 drawbar horsepower and 30.4 from the belt pulley. Rear wheels were 46 inches tall by a foot wide. The entire unit weighed 4,000 lbs. Total production was about 160,000 over a 30 year span.

The illustrations on this page, and the 3 following pages, were selected from the uncopyrighted 5th edition of John Deere's *Operation, Care & Repair of Farm Machinery*, circa 1930. Deere and Co. published this textbook to assist farm mechanics instructors in high schools and agricultural colleges throughout the nation, starting in 1920 and continuing for several decades.

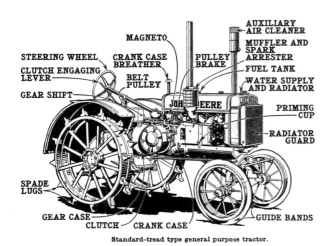
Standard-tread type general purpose tractor.

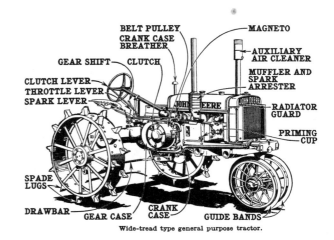
Wide-tread type general purpose tractor.

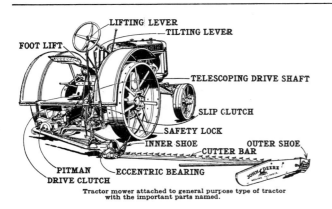
Tractor mower attached to general purpose type of tractor with the important parts named.

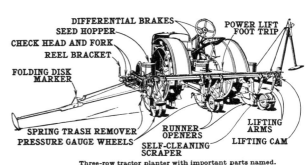
Three-row tractor planter with important parts named.

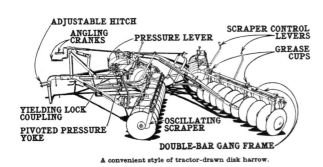
A convenient style of tractor-drawn disk harrow.

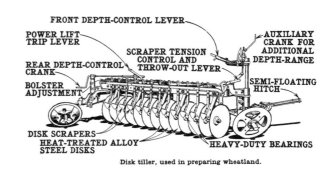
Disk tiller, used in preparing wheatland.

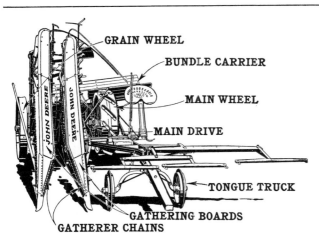
Corn binder with power-driven bundle carrier and tongue truck.

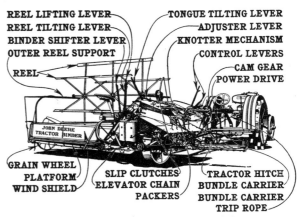
Tractor binder with tractor attached.

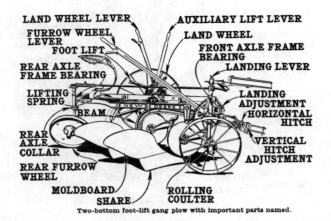
Two-bottom foot-lift gang plow with important parts named.

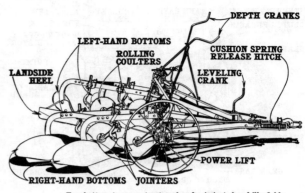
Two-bottom two-way tractor plow for irrigated or hilly fields.

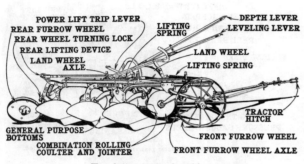
Three-bottom tractor plow with parts named.

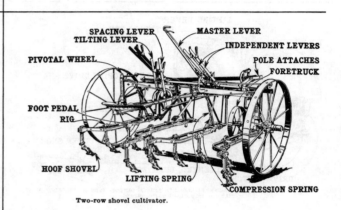
Two-row shovel cultivator.

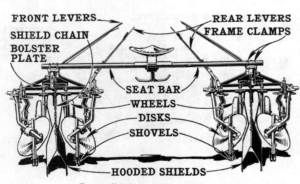
Two-row listed crop cultivator set for first cultivation.

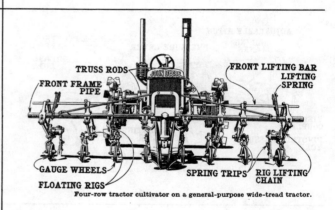
Four-row tractor cultivator on a general-purpose wide-tread tractor.

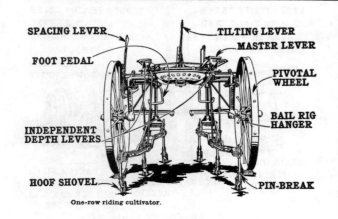
One-row riding cultivator.

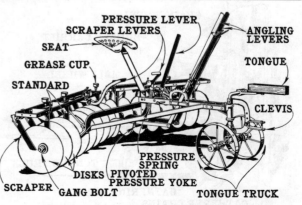
Horse-drawn disk harrow—this style may be converted to tractor.

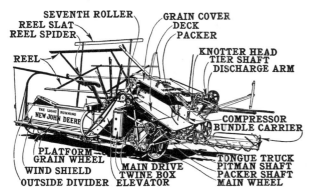
The grain binder is found on practically every farm.

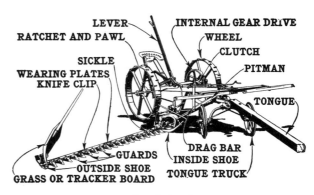
A popular type of mower.

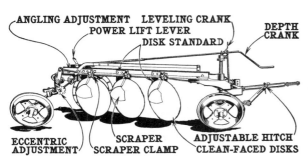
Tractor-drawn disk plow, important parts named.

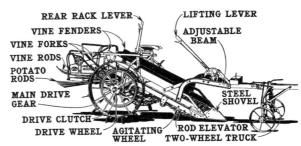
Elevator potato digger with important parts named.

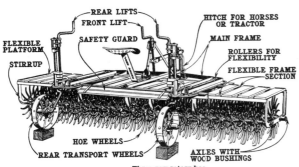
Three-row rotary hoe.

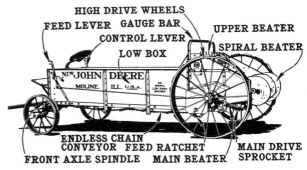
A tight-bottom type of manure spreader.

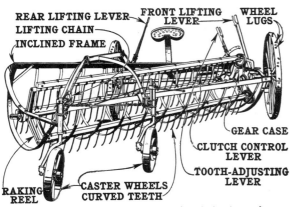
Side-delivery rake with the more important parts named.

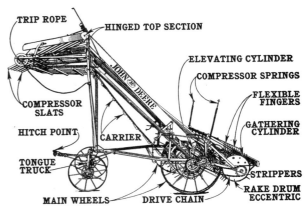
Double-cylinder loader with the more important parts named.

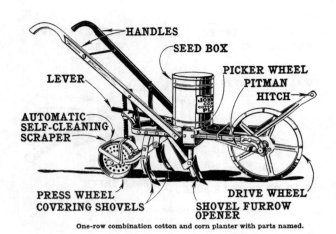
One-row combination cotton and corn planter with parts named.

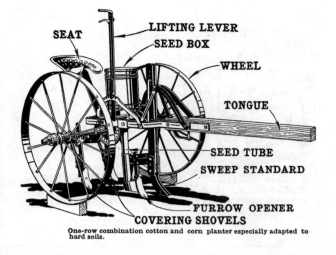
One-row combination cotton and corn planter especially adapted to hard soils.

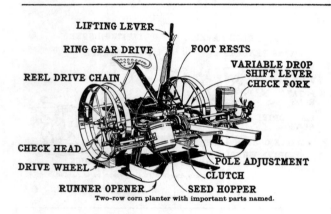
Two-row corn planter with important parts named.

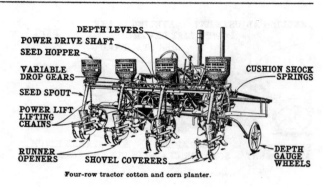
Four-row tractor cotton and corn planter.

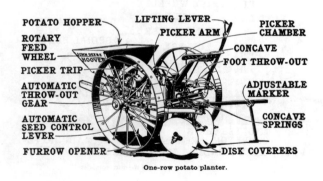
One-row potato planter.

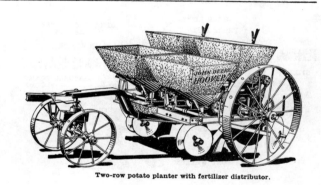
Two-row potato planter with fertilizer distributor.

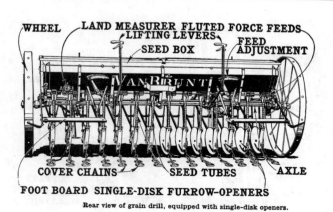
Rear view of grain drill, equipped with single-disk openers.

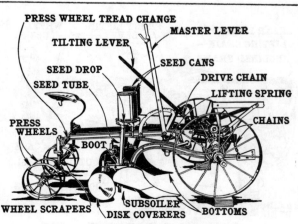
Two-row lister with important parts named.

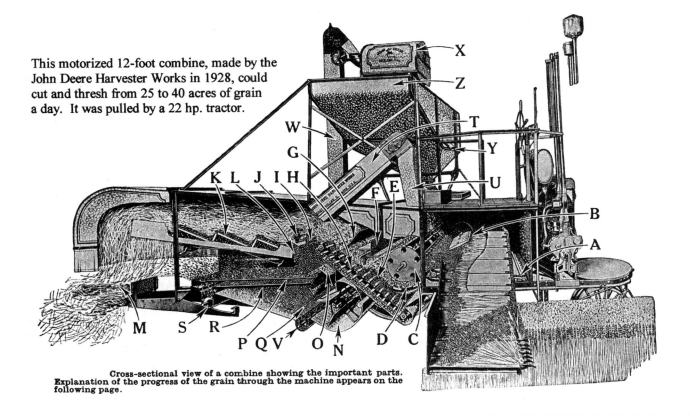

This motorized 12-foot combine, made by the John Deere Harvester Works in 1928, could cut and thresh from 25 to 40 acres of grain a day. It was pulled by a 22 hp. tractor.

Cross-sectional view of a combine showing the important parts. Explanation of the progress of the grain through the machine appears on the following page.

Cross Section of the Combine

The units are lettered so that the travel of grain and straw can be followed. The cut grain is carried by platform canvas to feeder house cross conveyor "A". This conveyor with aid of down beater "B" feeds grain to cylinder "C".

As grain travels between cylinder "C" and concaves "D" over finger grates "E" and back against beater behind cylinder "F", greater part of separation takes place. Beater "F" strips straw from cylinder and deflects grain and straw onto elevator behind cylinder. Most of threshed grain and chaff falls through bottom of concaves "D" and finger grate "E" onto elevator "H" behind cylinder. Grain and light chaff are deposited on raddle type carrier. Straw is carried on top of slats above threshed grain. Sectional retarders "G" assist beater in retarding straw in rapid movement and keep grain from being thrown "over".

As it leaves elevator, straw is agitated by picker "I" and bar beater "J" and is passed onto straw walkers "K" where it is pitched and tossed toward rear. Grain in straw falls through openings in walkers and flows back to front of shoe through grain return pans "L". Straw is then tossed out onto spreader "M".

As grain and chaff reach upper end of elevator "H" and as the straw is being worked on by picker "I" and bar-beater "J", it is subjected to blast from large undershot fan "N", through port "O", and the grain falls to separating shoe chaffer "P".

Another blast from fan "N" is directed at "Q" against chaffer "P" and lower sieve "R". This with aid of sieve agitation, blows chaff away and moves tailings to tailings auger "S". This auger carries them to tailings elevator "T" which conveys them into return spout "U" and back into feeder house.

In tailings elevator "T" is sieve which lets any clean grain through into return spout which carries it to carrier behind cylinder.

Clean grain, after dropping through chaffer "P" and sieve "R" is carried by clean grain auger "V" to elevator "W" on opposite side of machine, which delivers to rotary weed screen "X" on top of grain tank "Z".

129

Sears, Roebuck & Co., 1903

KENWOOD 3-HORSE POWER GASOLINE ENGINE

$98.50 Buys this High Grade Full 3-Horse Power Gasoline Engine, complete as described below, ready to operate, mounted on skids, crated and delivered free on board cars at Chicago.

THIS ENGINE IS A MODEL of perfection and completeness. It is built on practical lines, combining simplicity, durability and compactness, and developing the full amount of power we claim for it. The engine is complete in itself, requiring no extras to bring up its cost, and when it is received the only thing one has to do is to remove the crate, fill the water and gasoline tanks, and the engine is ready to run.

THIS ENGINE IS MADE ESPECIALLY FOR US by engine makers who have been in the gasoline engine business for several years, consequently it is not an experiment, but an up to date engine, containing every improvement which years of practical experience suggests. It is not a lightweight plaything weighing but a few hundred pounds, but is a good substantial engine weighing 1,000 pounds, and is suitable for running any kind of machinery which requires three-horse power or less. It is particularly adapted for small shops, farm use and for pumping water with either a windmill pump or a centrifugal pump, making a splendid machine for irrigation purposes. Is also a splendid engine for small electric light plants, and will easily drive a 30 to 35-light dynamo.

THE BASE IS A HEAVY CAST IRON BOX, weighing 250 pounds, is 4 feet long, 16 inches wide, 12 inches deep, divided and strengthened by partitions. In one end of this base is a compartment in which the 5-gallon galvanized steel gasoline tank is placed; the balance of the base serves as a water tank and holds 35 gallons. The base is covered with a heavy oak top, upon which the engine sets, the engine being securely bolted to the iron base.

THE ENGINE CYLINDER is five inches inside diameter, and piston makes a stroke of eight inches. Engine cylinder, engine head and valve are all surrounded by a very large water space through which the water is forced all the time the engine is running. The crank or main shaft is 1⅝ inches diameter, made of forged steel nicely turned and finished, and runs in extra long babbitted bearings in which there is ample provision for oiling and for taking up wear. Connecting rod is made of crucible steel and has brass bearings at both ends. Piston head is extra long,

THIS ENGINE IS A FOUR-CYCLE TYPE making one explosion, when necessary, to every second revolution of the crank shaft. It occupies a floor space of 48x28 inches, is 48 inches from bottom of base to top of cylinder, or 62 inches to top of muffler. Regular pulley is 10 inches diameter, with 6-inch straight face for 3-inch belt. Can be furnished with any diameter of 6-inch face pulley from 6 inches to 24 inches, but for pulley over 14 inches diameter an extra charge will be made. Engine is mounted on skids as shown in illustrations.

In this circa 1900 photo a salesman demonstrates his company's new gasoline-powered fodder/shredder/blower to a group of skeptical midwestern farmers.

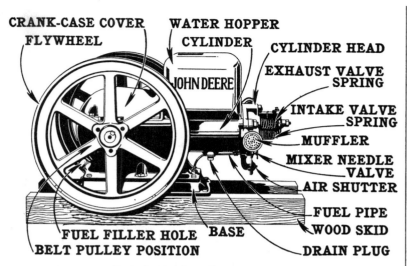

Typical style of gasoline engine

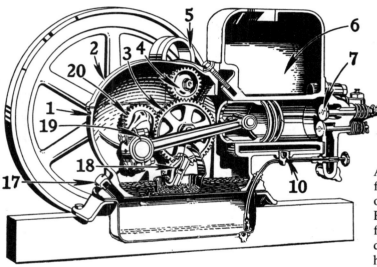

"R & V" GASOLINE ENGINES.

Triumph Horizontal Engines.
This is an exceedingly practical engine for farm use. It is a light, dependable and economical source of power. Hopper cooled it does away with air-cooled engine troubles. All "R & V" Engines are guaranteed to be economical. The Triumph is made in three sizes, one, two and four-horse power.

Volume-Governing Engine.
This is a strictly first-class motor, built for all general power purposes. The amount of fuel used at each charge is controlled by a governor. This gives smoother running and closer regulation. Governor is provided with a spring load which can be adjusted to give a speed variation of thirty per cent. to engine while in operation. Sizes of this machine vary from four to twenty-eight-horse power. As shown below, it is furnished mounted, if so desired.

Strictly High Grade for Heavy Farm Work.

Five Sizes. 6, 10, 12, 16 and 20-Hp.

Above (left) is a typical John Deere single-cylinder gasoline farm engine of the 1920's. On the right are several examples of Root and Vanderport engines distributed by Deere in 1910. Both types were typically mounted on wooden skids or oak-framed, iron-wheeled carts. Pin striped paint jobs and fancy decals completed the package. Today, these engines are highly collectable, and even have a monthly publication devoted to the hobby.

STATIONARY GAS ENGINES were displayed at the Paris Exposition of 1878, but they only developed "one man-power", the equivalent of one-twelfth horsepower. By 1895, several brands of gasoline and kerosene-burning farm engines were available. Montgomery Ward sold an "Electro Vapor" engine in sizes from 1/2 to 10 hp that ran on either natural gas or gasoline. In 1903, Sears & Roebuck offered a one-cylinder, three-horsepower, model for $98.50 that weighed one thousand pounds. It was used to pump water or power a 30-light electric generator.

By 1912, there were many brands of both water and air-cooled "hit and miss" governed engines on the market, developing from one to thirty-horsepower. International Harvester's best seller was an air-cooled, one-horsepower, single-cylinder, horizontal-style, called "The Tom Thumb Famous Engine". The Cushman Motor Works of Lincoln, Nebraska made a two-cylinder, eight-horsepower engine that weighed only 320 pounds. Rumely & Co. manufactured a kerosene oil burning engine called the "Oil Turn Motor" that developed a whopping 30 to 60 hp. A similar engine powered the Advance-Rumely "Oil Pull" tractor, which used cooling oil in its radiator instead of water.

The function of these various sized engines was to replace wind and animal power on the farm. They ran all kinds of machinery, including corn shellers, grinding mills, grain cleaners, feed cutters, silo fillers, hay balers, concrete mixers, water pumps, lighting plants, washing machines, butter churns; tool grinders, saws, lathes and post drills.

The obvious advantage of gas engines, over horses, was that you did not have to house and feed them when they were idle. A main disadvantage was that they sometimes refused to start during cold weather.

In the half-century between 1884 and 1934, at least a hundred companies were engaged in the gasoline and kerosene engine business. In addition to the previously mentioned firms there were colorful names such as: Alamo, Appleton, Bloomer, Caille, Detroit, Elgin, Flinchbaugh, Galloway, Gemmer, Gilson, Goetz, Jacobson, New Way, Peerless, Safety Vapor, Simple, Star, Stickney, Watkins, Webster, Worthington and Xargil.

When rural electrification arrived in the 1930's most gas-powered agricultural engines were replaced by smaller, cheaper, quieter, safer, and maintenance-free electric motors.

1892

CATALOGUE.

FRANK BROTHERS

DEALERS IN

AGRICULTURAL IMPLEMENTS

BUGGIES, ✹ CARTS, ✹ SPRING ✹ WAGONS,

Vehicles of Every Description, Etc.

33 AND 35 MAIN STREET,

San Francisco, Cal.

On the next 21 pages we have reproduced a portion of the Frank Brothers 1892 Catalog of Agricultural Implements, Wagons and Buggies. These illustrations provide the reader with a comparison of western goods vs. those popular on the east coast.

BUFORD WHEEL LANDSIDE SULKY.

THE LIGHTEST DRAFT PLOW MADE,

because *every pound of weight and pressure is carried on wheels*, a round-rimmed wheel being substituted for the ordinary bar landside, and running obliquely in the corner of the furrow, reduces the draft to the very lowest limit.

It does Superior Work in all Conditions of Soil and will plow the hardest ground without skip or jump. Any width and depth desired can be held uniformly. It will turn under out of sight heavy stalks, or other trash, in which a hand plow could not work.

It is Simple, Strong and Durable; constructed in a substantial manner, and of the best and most durable material, and is so simple any one can operate it perfectly.

The Power or Automatic Lift is very simple and never failing. The plow being carried out of the ground, by the team, by simply striking a pawl, which engages a hook with the wheel-clutch, while the team is in motion.

PRICE.

Complete, with double shin, patent soft center steel mould-board, extra share, wheel landside, castor cutter with chilled bearings, steel wheels, three-horse evener, neck yoke, tool box and power lift.

14 or 16 inches cut $65 00

Rock Island Clipper Plows.

(Formerly BUFORD CLIPPER.)

PRICE.

Double Shin. Medium Lands.

C. No. 6, cuts 14 in. $17 00
C. No. 7, " 12 in.   15 00
C. No. 7½, " 11 in. 14 50
C. No. 8, " 10 in. 13 50
   No. 8, " 9 in. 10 00
   No. 8½, " 8 in.   9 50

The land-bar and frog in these plows are welded as if in a solid piece, which gives as much strength as though share and landside were one, and not only allows the use of slip shares fitting sure and closely, but also assures the certainty of obtaining exact duplicates at any future period, by giving the number stamped on back.

The handle-brace is double riveted to the *standard*, and securely bolted to the *land-bar*, forming in connection with the handle a truss-brace, from the rear of the beam to the landside, adding great strength to that part of the plow which must endure greatest strain, and at the same time raising the lower part of the handle so it cannot accumulate dirt and trash.

The shares all have a heavy, solid piece of steel welded to the point. Double Shinned Plows have a solid piece of steel welded to the soft center molds on upper surface, thus adding about three times its original capacity to that part which is first to wear.

The success of these plows is largely due to the care exercised in their construction, to secure uniform shape, thorough and even hardness, and a fine polished finish, giving the qualities for which they are noted and unsurpassed, of "easy draft," "level running," and "scouring in any soil."

NOTE THE SUPERIOR CONSTRUCTION OF THESE PLOWS, AS SHOWN ON OPPOSITE PAGE.

THE ELI GANG PLOW.

FOUR PLOW, 10-IN. GANG.

GREAT ADVANTAGES. The weights of Plow and furrow slices are carried on wheels, because the patent lifting device makes it possible to handle them. On other plows it is difficult to handle the levers, unless made so long that little action is obtained; but on the Eli it is unnecessary to have levers as long as others, because of the perfect way in which the draft balances the weight and "suck" of plow, therefore much greater movement is obtained on the crank of axle, and the plow can be raised higher from the ground. This is especially true, as the axles are longer than others, so that the points of the plows are well raised from the surface, when not in use. A uniform depth of plowing is insured at all times by having wheels properly placed.

The patents on this Plow are in the "joint, or knuckles," through which levers regulate the Plow. By this device an independent adjustment of each axle is secured, and at the same time complete control of the Plow by one lever, while at work; a valuable feature in opening and closing lands and on side hills.

The device for raising Plow is simple and effective. It consists of a draft rod drawing from a pendant arm on the axle, to which team is hitched, perfectly balancing weight of Plow, the operator, and the tendency of the Plow to run into the ground, and enabling operator to raise or lower it with the least exertion. So easy is it to handle, that a child with strength enough to draw the spring bolt from the ratchet can control the depth of plowing, and raise the plow from the ground. No other plows have this device.

PRICE WITH IMPROVED LAND GAUGE AND EXTRA SHARES.

Four Gang, 10-in.	$90 00	Two Gang, 14-in., Walking	$70 00
Four Gang, 8-in.	85 00	Two Gang, 12-in., Walking	65 00
Three Gang, 10-in.	80 00	Two Gang, 10-in., Walking	60 00
Three Gang, 8-in.	75 00	Rolling Coulters, each extra	5 00
Riding attachments for Two and Three Gangs			$10 00
Pole attachment for Two Plow Gangs			4 00
Eveners for Four Horses Abreast			7 00

Beams, Levers and Axles are so arranged that one Plow may be removed, reducing a 4 Plow to a 3 Plow Gang, or a 3 Plow to a 2 Plow.

THREE PLOW, 10-IN. GANG.

ELI GANGS AND SULKIES.

TWO PLOW, TURF AND STUBBLE BOTTOMS, COULTERS EXTRA.

The Eli Sulky, shown below, is a Plow which for general utility cannot be excelled. It is a perfect plow for a walker, and requires no change for riding. It is a perfect riding plow, and requires no change for walking. The levers controlling are handled with equal ease either way. The point raises high when out of the ground. The wheels are large; weight of plow, driver and furrow slice are carried on them, and draft is therefore necessarily light. The *spring* which is attached to land axle, on single and double plows, permits the wheel to rise and fall without lifting plow while passing over rough or ridged ground, and thus keeps a level and even furrow. A uniform depth of plowing is insured at all times by having wheels located opposite the point of plow. It is provided with lock on rear wheel so that plow may be prevented from running on to horses when on the road. Its second lever, aside from enabling easy leveling of plow, assists very materially in maintaining a perfectly straight "land." The riding attachment wheel is swiveled, and has an automatic lock to hold it rigid, the bolt of which is released when turning by a foot lever, conveniently located. This permits turning perfectly square corners either to the right or left.

PRICE.

Land Gauge, Double Shin, Soft Center Steel Mold, Anti-Friction Rolling Cutter, Three-Horse Equalizer, Chilled Hub Boxes, Medium Landside, Extra Share and Turf and Stubble Bottom.

14 or 16-in. cut, Eli Sulky **$60 00**
Without Riding Attachment, deduct **10 00**

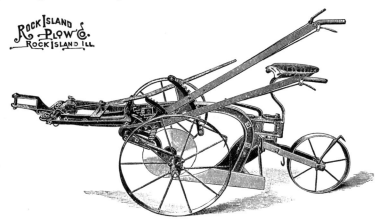

"ELI" Perfect as a Walker.
Complete as a Sulky.

THE McLEAN CULTIVATORS.

No. 1. Cultivator. 7 Shovels—8 inches space between Standards.

MANUFACTURED BY

Jensen & Lauritzen, Watsonville, Cal.

The McLean Cultivator will do a greater variety of work than any Cultivator on the Pacific Coast. It is designed to supplant the plow in orchards and summer crop land. It cuts deep, effectually roots out weeds, and does more and better work than a plow. It has been adopted as the best machine of its class, by the farmers and orchardists of Pajaro Valley, by the orchardists and vineyardists of the Santa Clara Valley, by the bean farmers of Santa Barbara, and by the orange growers of Riverside.

Meeting with the endorsement of the farmers and fruit raisers of so many different sections of the State, and receiving first premiums wherever exhibited, are strong grounds for the claim that the McLean Cultivator has no equal. Many improvements for 1892, including malleable interchangeable hub boxes.

No. 4 Cultivator has 11 teeth with 6 in. space between standards. It is made strong, is well braced and adapted for heavy work. 6 ft. cut.

PRICE.

No. 1.	McLean Cultivator, 2 Horse, 7 Tooth				$ 50 00
No. 3.	"	"	3 " 9 "		55 00
No. 4.	"	"	4 " 11 "		75 00

Shovels, each, 75 cents; chisels, each, 40 cents.

The McSherry Grain Drill,

No. 3.

Adjustable Force Feed without change of Cog-Wheels.

The McSherry combines everything the most successful and practical farmers want, and is beyond question the right and best machine for farmers to use.

The flow of seed is regular and it cannot bunch, break or crack the grain. It can be regulated in a moment to sow any desired quantity per acre. The feed cannot choke and works equally well on hilly or level land and when driven fast or slow. The wheels are of the best quality of seasoned timber, with wide tires and cold rolled iron axles.

It has long hoes of improved shape working in any soil. They can be worked straight or zig-zag and raised by a lever at the side or held to their work with a spring by means of the same lever.

Each Drill has a Grass Seed Attachment and Land Measure; is properly and evenly balanced and light draft and very easy running.

The McSherry has been sold in California for the past 15 years and is everywhere considered the best Drill made. It is greatly improved for this season and its use is a sure investment, bound to give you properly seeded land and the best crop possible.

PRICES OF THE NEW McSHERRY GRAIN DRILL, No. 3.

9 Hoe,	6 or 7 inches	$ 90 00
12 "	6 or 7 "	110 00
15 "	6 or 7 "	140 00
18 "	6 or 7 "	165 00
20 "	7 "	180 00
8 Feet Broad Cast Seeder		85 00

THE FAMOUS LOCK-LEVER SULKY RAKE,

Is Operated by Driver's Weight, and has no Complicated Horse Machinery.

The Famous is a simple, strong Rake, with an immense capacity and great durability. Its simplicity commends it to all. It is made of thoroughly seasoned hard wood. The axle arm is of 1½ round iron, securely fastened to the wood head, and bent in such a way that the center of the axle arm is the center of the axle head. The shafts being hinged to the upper part of the axle, it follows that the weight of the driver discharges the load as soon as the lock-lever is disengaged.

The great feature of this Rake is its **Lock-Lever**, which has an ingeniously contrived joint that holds the Rake firmly to its work, when down, and relieves the driver from all care in that direction. Discharging the Rake is effected by simply giving the slightest upward motion to the lever handle, which, unlocking it, causes the weight of the driver to tilt the Rake-Head and discharge its contents.

A Child can Operate the Famous with Ease.

The Seat Standard is of spring steel, is well supported, and to the seat are attached the rods that carry the over-cleaner attachment. The teeth are of crucible cast steel, have a double coil, and are securely fastened to the Rake-Head. The wheels are metal, 50 inches high. Every Famous Rake has a **Patent Spring-Seat**. Every Famous Rake gives satisfaction, and every Farmer ought to have a Famous Rake.

The Famous is a Standard Rake of Proven Merit.

PRICE.

With Steel Wheels	$30 00
Extra Teeth, per dozen	6 00

Furnished with Shafts.

WALTER A. WOOD ENCLOSED GEAR MOWER.

Four widths of Cut, with Tilting Lever. The merit of the Walter A. Wood mower causes it to take first premium on all competitive occasions and at all the great WORLD'S FAIRS it has received highest honors.

Among the causes of this supremacy notice: Strength, lightness and simplicity; highest class steels, malleables and brasses; direct under-draft by the floating droop-frame; only purely floating droop-frame and cutters; steadiness of machine and pole *when at work;* wheels firm to ground, whether going fast or slow; both wheels driving with full power; wide frames for the wide cuts; no neck weight or side draft; freedom from "choking" in the grass; firm but free and adjustable pitman connection.

Walter A. Wood's mower-gearing is celebrated everywhere. Some suppose this gearing works so smooth and wears so long because of its special arrangement, but that is not the only reason—it is largely owing to workmanship, and the quality of metal used. Its appearance can be imitated, but not its quality. Move it with the hand and notice how perfectly it is fitted and how easily it runs. Notice, also, that there is only one bearing besides the main axle of the machine (which is made to serve for a bearing to one of the gears). Notice that the gear which turns on the main axle does not have to turn fast on its bearing, as it turns in the same direction with the axle. The bearing turns in the gear fast enough to save considerable in the way of friction.

Wonderful perfection has been reached in Walter A. Wood's cutting apparatus. There is no cutlery better made for use. There is no cutting apparatus that gives so little trouble, lasts so long and runs so freely. The sections are made at their own works of the best cutlery steel. The guards, made of their own special mixture of malleable iron, are milled all alike to fit the bar—the only milled guards made—and are armed with steel guard-plates of the same fine steel and temper as the sections. The cutting apparatus contributes not a little to the never failing victory of the Wood in all draft contests.

In all Walter A. Wood's mowers he has insisted on not suspending any weight on the neckyoke of the team. Every pound of the machine the team carries on the neckyoke subtracts from the grip of the driving wheels on the ground. Argument will not make the wheels hug the ground. Nothing will do it but the law of gravitation. Therefore the driving-wheels should have the benefit of all the weight that can be concentrated on them. Again, every pound of the machine that the team holds up by the neckyoke makes wear and tear for their shoulders and fore legs. Give your horses a chance to step freely with their fore legs and pull with their hind legs. Do not load them down in front.

Buy the Wood Mower.

THE THOMAS SULKY RAKE.

STRONG, DURABLE, AND A COMFORT TO THE RANCHER.

The Thomas is sold on its merits; all we ask of any one in search of a *good rake*, one that any child of ten can operate as easily as a man—is to *get into the seat and operate it*, examine the principle upon which it is built, its many original and distinctive points of superiority, and he must admit that it is, indeed, one of **the Best Rakes in the Market.**

Its largest sale is where it is best known, and where one goes "there's more to follow." We never knew a dealer to fail in securing and holding a large trade on this rake, at fair paying prices, who was *familiar* with its many good points, and took the pains to impress them properly upon his customers.

It is *the* rake for the farmer to buy, or the dealer to handle. It rarely requires repairs; it gives universal satisfaction; it will rake corn stalks as well as hay, and we have heard of it being used to rake hedge trimmings, which is conclusive evidence of its great strength.

WHY THE THOMAS EXCELS.

Because it is the easiest dumping rake.

Because a long brass spring is over each tooth. This enables the teeth to adjust themselves to the ground, causing less scratching and breakage.

Because its overhanging cleaner, with rollers on to prevent friction with teeth, turns the hay in dumping, brings the green hay on top, leaving it a loose windrow in good condition for drying.

Because it makes the largest windrows, for use in bunching hay alone, is worth more than the difference in price between it and other rakes. Teeth are shaped specially for taking up hay, and leaving dirt and trash.

Because nearly all iron used is either malleable or wrought, every part is *strong and durable*.

Because it is *comfortable and convenient to the operator*.

Because adapted for all kinds of work and ground.

PRICE.

Thomas Rake, Steel Wheels...$35 00
Thomas Rake, Wood Wheels... 32 50

Buford Combined Riding and Walking Cultivator.

Buford's Combined Riding and Walking Cultivator...................................$50 00
Fifth Tooth Attachment for above... 5 00

The special features of the Cultivator are: It is practically all Iron and Steel, having no wood except the tongue, evener bar, singletrees and handles.

The axles can be extended or contracted at will, to accommodate different widths of rows, adapting it to the cultivation of all kinds of crops planted in rows.

The couplings are of malleable iron, and have long bearings, both vertical and lateral, insuring firm and steady work. They also regulate the width between the beams by an adjustment on the axle sleeve.

The sleeves to which the shovels are attached, are so made that the shovels can be lowered without changing their pitch.

The shovels are made on the only true principle; they will throw the soil to and from the plant without crowding in the least, and being thoroughly hardened, will scour in any soil.

"Iron Age" Cultivator

5-Tooth, Iron Age Cultivator; weight, 60 lbs.......................................$ 8 50
7-Tooth, Iron Age Cultivator; weight, 65 lbs....................................... 11 00
Horse Hoe Attachment for 5-Tooth.. 2 00
Without Wheel, deduct... 1 00

A neat, light, strong, five tooth, iron frame, adjustable Cultivator, and as complete and perfect a tool of its class as is to be found in the market. All of the shanks are adjustable to set shovels at different angles; the beams can be contracted or expanded to any required width, adapting it to the cultivation of all kinds of small crops. The shovels are made of cast-steel, and are reversible. Furnished either plain or with wheel; also furnished with hoes and sweeps if so ordered.

FRANK BROTHERS, 33 & 35 MAIN STREET,

HENNEY BUGGY CO. VEHICLES.

DESCRIPTION OF MATERIAL USED IN THIS WORK.

Wheels—Best second growth, Banded Wood Hub, or Best Sarven Patent, Rims selected and screwed on each side of spoke-tenons to prevent splitting. Round edge, steel Tires.

Gearing—Choice selected, thoroughly seasoned hickory.

Bodies—First-class selected yellow poplar, and second growth white ash, well glued, screwed and plugged.

Axles—Solid steel, manufactured expressly for this work, double collar on the back with a chamber between, called a sand box, to stop particles of sand from getting on the arm of axle, or oil from getting outside.

Springs—Finest quality, oil-tempered graded steel springs, fully warranted.

Iron Work—All Bolts, Clips and Forgings best Norway Iron.

Tops—Are made of the best brands of *Hand Buffed* Leather; are lined with finest English cloth, either blue or green, with second growth bows, leather covered.

Trimmings—Fine English dark green or blue wool-dyed broadcloth, or dark green or brown trimming leather.

RUGS—Best Wilton or Velvet Carpet, Leather Boots and Panel-Toe Pads.

Dashes and Fenders—*All solid foot*, covered with No. 1 dash leather, well stitched.

Painting—*Only the best* paints and varnishes that can be purchased are used, and none but skilled workmen employed in this as well as all other departments.

BODIES—Finely finished in black, and GEARINGS *in rich dark green, with double fine line glazed stripe.*

DIMENSIONS.

Regular Sized Piano Bodies—50 in. long, 25 in. wide, 8 in. panel.
Three-Quarter size Piano Body—50 in. long, 22 in. wide, 8 in. panel.
Buggy Wheels—3 ft. 6 in. front, 3 ft. 10 in. hind.
Phaeton Wheels—3 ft. 2 in. front, 3 ft. 10 in. hind.
Axles—*Regular Buggy*—7/8 x 15-16.
Three-quarter size—3/4 x 13-16, 4 ft. 4 in. track. Wrought boxes, finely fitted.
Surrey Axles—1 x 1 1-16.

NOTE.—Standard Track for light Carriages, as adopted by the Carriage Builders' National Association, is 4 ft. 8 in. out to out.

SECHLER & CO.'S BUGGIES.

Sechler & Co. Grade or "S. & Co." Body, Seat and Gear, well made of thoroughly seasoned lumber. Gearing, double collar steel axle (swaged) fan-tail, genuine Sarven wheels, oil tempered springs. Top, Sechler & Co. "Perfection" four steel-bow, corded, leather quarters and back stays, rubber drill roof and curtains, Brewster fasteners; upholstering, cloth or leather. Well painted and finished throughout. Half carpet, apron and boot, or full carpet, boot extra.

Double Straight Perches are used on all Buggies except Concord and Timken Springs. Sechler & Co. "Perfection" Fifth Wheels on all except Concord and Timken Spring Buggies.

OUR WARRANTY.

We warrant all our vehicles to be of good material and workmanship, well made in every particular. Should any breakage occur within one year from date of purchase from us, by reason of defective material or workmanship, repairs for same will be furnished free of charge, upon the purchaser producing the broken or defective parts as evidence, unless a satisfactory price is agreed upon with us for said repairs, the amount of which shall be paid in cash.

We do not engage to make good all wheels, axles or springs that may be broken during the first year's use of the vehicle, but only such as show defect in the material.

In all cases where breakages occur, the broken parts must be furnished to us or our agents for inspection, before any claim can be allowed.

This warranty is not to be constructed to mean that any purchaser of a vehicle from us is entitled to repairs on it when the broken parts show that the timber or iron is good and the workmanship faultless.

Frank Brothers.

No. 27. Flint Perfection Wagon.

No. 27, with shafts, $70 00. Brake, $10 00. Canopy Top, $20 00. Pole, $12 00.

Body, 6 ft. in length, 29 inches wide, with 7 in. panel. The **Gear** is a combination of half elliptic spring at the rear end, suspended directly over the axle, and ¼ elliptic springs in front, attached to head block, intended to **prevent any pitching of the load.** **Wheels,** 1 in. Sarven, 3 ft. 6 in. front, 3 ft. 10 in. hind, with 1x3-16 round edge *Steel Tire*. **Axles,** double collar steel. **Painting,** Body, Black; Gear, Brewster Green. **Upholstering,** Corduroy, Rubber, Evans Leather or Tannette. Seats are moveable. Weight, 350 lbs. Capacity, 600 lbs.

This Wagon rides perfectly with two or four passengers.

No. 29. Flint Three-Spring Wagon.

No. 29, with shafts, $80 00. Brake, $10 00. Canopy Top, $20 00. Pole, $12 00.

Body, 6½ ft. in length, with drop end-gate; panel, 8 in. deep, with hardwood rail. **Gear** has double reach, ironed full length; 12 in. McCabe 5th wheel, with king bolt in rear of axle. **Springs,** 1¼ in., 4 plate. **Axles,** 1 1-16 in. square. **Wheels,** Sarven, with heavy hub, spoke and felloe, 3 ft. 6 in. and 3 ft. 10 in., with ¼x1 in. round edge steel tire. **Upholstering,** Corduroy, Evans Leather or Tannette. Weight, 500 lbs. Capacity, 800 lbs.

This wagon is built with **Special Corner Irons and Canopy Top Fasteners Combined,** which greatly strengthen the body.

No. 130. Three-Spring Express Wagon with Panel Body.

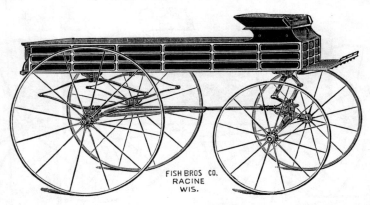

Gears ironed with king bolt in front of axle and head block, making the best coupling known. Wheels 3 feet 3 inches and 4 feet 3 inches high. Price includes shafts with steps attached. Brake.

No.	Axles.	Weight.	Capacity.	Body.	Price.
130 A	1⅛ in.	540 lbs.	1,000 lbs.	6 ft. 6 in. x 3 ft. 7 in. x 10 in.	$140 00
130 B	1¼ "	640 "	1,200 "	7 " x 3 ft. 7 in. x 10 in.	155 00

No. 132. Combination Spring Express Wagon.

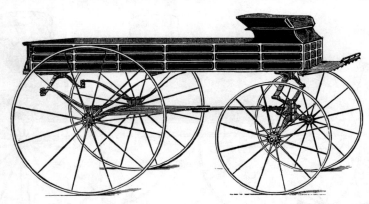

All our Express wagons have drop end-gates. Sarven patent wheels. Steel tire. Oil-tempered springs. Gears ironed with king bolt in front of axle and head block, the best and strongest coupling made. Priced with shafts only. Pole, extra, $15.00. Brake. Painted, red gear, green body.

No.	Axles.	Weight.	Capacity.	Body.	Wheels.	Price.
132 A	1⅛ in.	540 lbs.	1,000 lbs.	7 ft. 6 in. x 3 ft. 7 in. x 10 in.	3 ft. 3 and 4 ft. 3	$150 00
132 B	1¼ "	650 "	1,200 "	7 " 6 " x 3 " 7 " x 10 "	3 " 3 " 4 " 3	165 00

SAN FRANCISCO, CAL.

No. 133. Low Three-Spring Light Delivery Wagon.

This body is 7 feet 6 inches x 3 feet x 8 inches and has raised panels, as shown in cut. The body is only 33 inches from the ground. The rear springs are hung under the axle, and the front axle is dropped so as to get the wagon as low as the crank axle wagons so much in use, besides being stronger, neater and more desirable for a light low wagon. There is none better. Priced with Shafts and Brake. Painted, gear red; body green.

No.	Axles.	Tires.	Wheels.	Weight.	Capacity.	Price.
133 A	1⅛ in.	1 in. cr 1⅛ in.	3 ft. 6 in. x 4 ft.	575 lbs.	1,000 lbs.	$140 00
133 B	1¼ "	1⅛ " " 1¼ "	3 " 6 " x 4 "	600 "	1,200 "	155 00

No. 240. Our Boss Express Wagon.

This wagon is made with gear similar to No. 208 Boss Wagon, only ironed heavier for delivery work. It is made with Two-Panel Bed and is a very popular job for light express delivery work. Body, 6 feet 6 inches x 3 feet x 8 inches. Wheels, 3 feet 6 inches and 3 feet 10 inches. Furnished with Express Shafts with steps.

No. 240, 1⅛ inch Axle; Weight, 375 lbs.; Capacity, 800 lbs. Price.................. $115 00
Brake, additional.. 10 00

No. 40. FLINT SPINDLE WAGON.

No. 40. Side Spring Spindle Wagon..$70 00

FLINT ADJUSTABLE POLE.

Excels In the simplicity and strength of adjustment. Fits any width vehicle from 32 to 49 inches. Self-centering when attached. Guaranteed not to rattle. Is no experiment, having been upon the market and subjected to the severest tests for nine years. A trial will convince the most skeptical that it is thoroughly practical and a specialty worthy of its enormous sale. Now within the reach of all, the price having been reduced to that of a common pole.

DESCRIPTION.

The braces are securely fastened to cross bar, which moves forward and backward upon the plates clipped to under side of pole. Both plates are grooved, so that by tightening the nut in any position the fastening is perfect and secure. Circles are socketed at rear end of pole, allowing whatever spread is necessary. Nothing but the very best material is used in the construction of these poles.

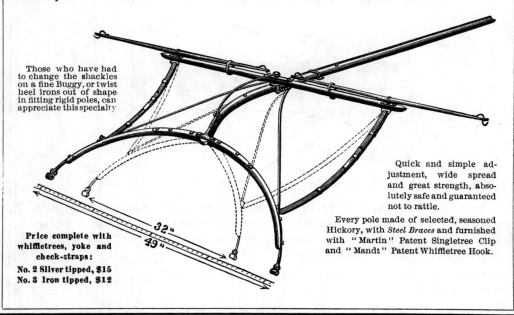

Those who have had to change the shackles on a fine Buggy, or twist heel irons out of shape in fitting rigid poles, can appreciate this specialty

Price complete with whiffletrees, yoke and check-straps:
No. 2 Silver tipped, $15
No. 3 Iron tipped, $12

Quick and simple adjustment, wide spread and great strength, absolutely safe and guaranteed not to rattle.

Every pole made of selected, seasoned Hickory, with *Steel Braces* and furnished with "Martin" Patent Singletree Clip and "Mandt" Patent Whiffletree Hook.

California Four-Spring Wagon.

WITH THREE SEATS AND FULL LAZY BACKS.

Has plain panels with square ends and is especially suitable for stage, excursion and general passenger service. Similar in construction to No. 147, but proportionately stronger. Has extra wide seats affording ample room for nine passengers.

1⅜ Inch Axle, Three Seats and Brake	$200 00
Trunk Rack for same	10 00

No. 161. Our New One-Seat Western Buckboard.

Made strong for rough mountain roads. Springs under front of body to relieve the jar from feet such as is common in all Buckboards without springs. Corduroy trimming. Painted, red gear; black body. Price includes Shafts and Brake. Pole, extra.

No. 161 A, Axle 1⅛ in.; Weight, 390 pounds; Body, 7 ft. x 2 ft. 6 in.	$100 00
Pole	12 00

No. 36. SPINDLE WAGON. THREE-QUARTER SIZE.

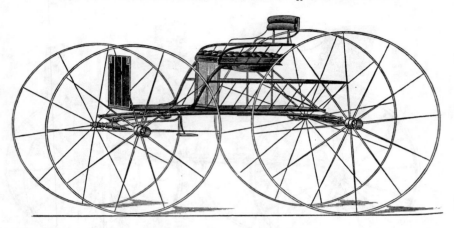

No. 36.   Henney, Standard Grade, Shafts... $125 00

Built with great care for speeding and road purposes.

"HENNEY STEEL GEAR" ROAD WAGON.

Has springs of the very finest quality, oil tempered, 55 inches long, hence a soft and easy motion is secured. There is no side swaying to the body, and **the Patent Truss Brace,** connecting the springs, insures the perfect track of the wagon under all circumstances.

The longitudinal arrangement of the springs connecting the axles, and the absence of the usual **rigid reach,** gives to this **ALL STEEL GEARING a WITHEY ENDURANCE** not to be had in any other similar gear, in fact, it must be literally **WORN OUT** before it can be said to **REQUIRE REPAIRS.**

The **"HENNEY STEEL GEAR"** is a perfect structure. No turning of axles when springs are depressed. No wood to shrink or decay, and the cross-braces insure under all circumstances the same **Unerring Track of the Wheels,** as if its axles were connected by a **Rigid Perch** and **Stays.**

Furnished ordinarily without top, but if desired top can be supplied as shown below. This same gear is also built as No. 84, with Concord body.

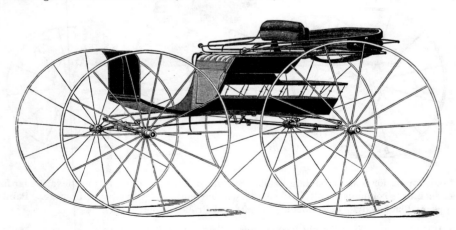

No. 81.   "Henney Steel Gear" Road Wagon, Spindle Body, Open, Shafts .... $ 75 00
No. 81 A.   Same, with Quarter Leather Buggy Top................................. 100 00

No. 33. Four-Passenger, Open, End Spring, Piano Box

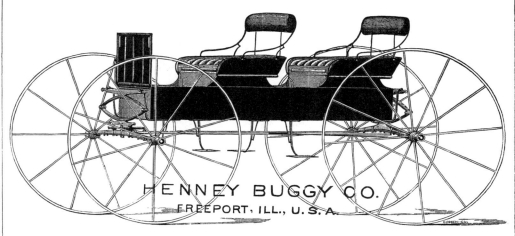

No. 33. Henney Piano Box Wagon, Open, Pole, $125 00.
Same with Canopy Top, $150 00.

No. 30. Four-Passenger Brewster, with Canopy Top.

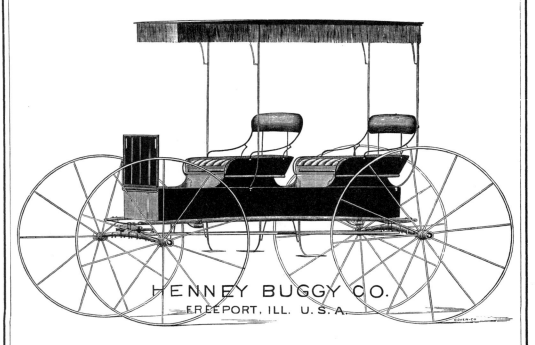

No. 30. Henney Brewster Wagon, Canopy Top, Pole, $150 00.
Same without Top, $125 00.

No. 121. SECHLER BREWSTER SIDE BAR BUGGY.

With Sechler Perfection Top, see page 41.

No. 121. Brewster Piano Box Buggy, with Leather Quarter Top and Rubber Side Curtains $115 00
No. 121 A. Same, with Full Leather Top and Rubber Side Curtains.................. 125 00
No. 121 B. Brewster Piano Box Open Buggy.. 90 00
1⅛ inch axle, additional, $5 00.   Pole, $12 00.   Shafts, $7 50.

The Dexter Queen, shown below, is "A Combination of Two Top Parallel Springs, Two Bottom Diagonal Springs and a Center Stay or Body Brace." This combination secures strength with elasticity. The springs lying in the direction of travel are a substitute both for reach and side bar, and furnish that most desirable substitute, an elastic reach, diagonal from king bolt, toward rear spindles, keeping a perfect track, and adding greatly to the safety and durability of the vehicle.

No. 42. SECHLER CORNING BODY BUGGY.

With Sechler Perfection Top, see page 41.
No. 42. End Spring Corning Body Buggy.   Prices same as for No. 48 with Piano Box.

No. 147. California Four-Spring Wagon.

Our Fish Brothers Wagons are especially built for California trade, and hard service. They have drop end-gates, are Full Leather Trimmed, Plain Bodies with Bracket Fronts, and are wide enough to take fruit and butter boxes. Full backs on seats. Best Double Collar Steel Axles and Oil Tempered Springs, Clipped King Bolt, Wheels A Grade, with Steel Tires. Either inside or outside brake. Pole, Neckyoke and Doubletrees complete.

No. 147, 1⅛ Inch Axle, Two Seats and Brake...$150 00
No. 147, 1¼ "   "   "   "   " 160 00
For Standing Top as shown below add... 35 00

No. 147 A. California Four-Spring Wagon with Top.

FISH BROTHERS CALIFORNIA WAGON.

EASTERN BOX BED.

Complete with Eastern Double Box Bed, California Roller Brake, Doubletrees, Neck Yoke, Stay Chains, etc.

THIMBLE SKEIN.					STEEL AXLE.				
No.	Size Skein.	Tire.	Capacity.	Price.	No.	Size Skein.	Tire.	Capacity.	Price.
B 29	2¾ x 8½	2 inch	2,000 lbs	$110 00	B 45	1⅜ x 8½	1½ in.	2,000 lbs	$120 00
B 31	3 x 9	2 "	2,600 "	115 00	B 47	1½ x 8½	2 "	3,000 "	125 00
B 33	3¼ x 10	2¼ "	4,000 "	130 00	B 49	1⅝ x 9½	2 "	4,000 "	135 00
B 35	3½ x 11	2½ "	4,800 "	140 00	B 51	1¾ x 9½	2¼ "	4,800 "	145 00
B 37	3¾ x 12	2½ "	6,000 "	160 00	B 53	1⅞ x 10½	2¼ "	6,000 "	155 00
B 39	4 x 12	3 "	7,000 "	175 00	B 55	2 x 10½	2½ "	7,000 "	165 00

FISH BROTHERS DEMOCRAT WAGON.

A light and convenient spring wagon for Ranch and Market use. Has *two* Spring Seats, Brake, Drop End-Gate, and either Pole or Shafts. When bed is taken off it leaves a handy running gear with bolster stakes that can be used for light work. Height of wheels 3 feet 8 inches and 4 feet 2 inches. Farm wagon style and finish.

No. 113.	2¼ inch Thimble Skein, Capacity 1,000 lbs., Weight 700 lbs.	$110 00
No. 115.	2½ inch Thimble Skein, Capacity 1,250 lbs., Weight 740 lbs.	115 00
No. 121.	1¼ inch Steel Axle, Capacity 1,500 lbs., Weight 760 lbs.	120 00
No. 123.	1⅜ inch Steel Axle, Capacity 2,000 lbs., Weight 810 lbs.	130 00

FISH BROTHERS CALIFORNIA WAGON.

STAKE RACK BED.

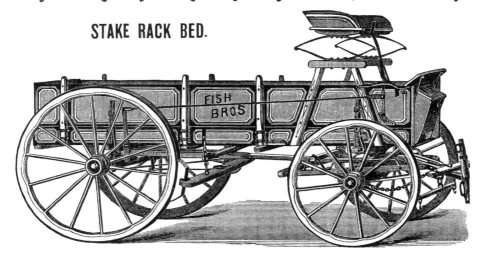

Complete with California Stake Rack Bed, California Roller Brake, Doubletrees, Neck Yoke, Stay Chains, Etc.

No stronger guarantee of merit in a wagon can be furnished than that it is of **FISH BROTHERS'** manufacture; this is especially true of their California Wagon, which is built to stand the trying demands of California usage.

Carefully made in the most thorough manner, the **FISH BROTHERS** Wagon is offered you with confidence in the manufacturers' claim that it is "The Best Wagon on Wheels."

Thimble Skein, with FISH BROTHERS Patent Self-Oiling Skeins.

Number.	Size Skein.	Size Tire.	Estimated Capacity.	Weight.	Price.
A 29	2¾ x 8½	2 inches	2,000	1025 lbs.	$ 120 00
A 31	3 x 9	2 "	2,600	1125 "	125 00
A 33	3¼ x 10	2¼ "	4,000	1420 "	140 00
A 35	3½ x 11	2½ "	4,800	1540 "	150 00
A 37	3¾ x 12	2½ "	6,000	1815 "	175 00
A 39	4 x 12	3 "	7,000	1945 "	190 00

No. 519. Spindle Body Wagon with Patent Boss Springs.

1 inch Axle. Weight, 275 lbs. Price, **$70.00.**

By using the new Boss Springs and crank axles, we are enabled to furnish a larger body, without lengthening the gear, thereby giving good room back of the seat, and making one of the best light business wagons offered. Gear red, body black. Axles, 1 inch. Tire, ⅞ x 3-16. Body, 4 feet 6½ x 26 inches. Shafts.

The following 48 pages are reprinted from Montgomery Ward's 1896 Agricultural Tool Catalog. Aaron Montgomery Ward was a former traveling salesman who had served the farming community for many years and knew exactly what farmers needed, and of their desire to elimate middlemen. Montgomery Ward Was founded in 1872 with $2,400 in capital.

A226—The Planet Jr. Combined Drill, Wheel Hoe, Cultivator, Rake and Plow.

This is unquestionably the most popular and perfect machine of its kind made. As a seed drill it is the same as A224 except in size; it holds one quart. It has all the tools shown in cut. All blades are steel, tempered and polished. The Rakes are invaluable in preparing the ground for planting, for covering seeds, first cultivation, etc. The Hoes work closely and safely all rows up to 16-inches wide at one passage, leaving the ground nearly level. The plow opens furrows, covers them, hills, plows to and from, etc. The Cultivator teeth are admirably adapted to deep mellowing of the soil and marking out. Taken as a whole, this combined tool is the nearest approach to perfection for the use of a gardener that can well be devised in a single implement. Weight, 40 lbs...........Price, $9.00

A232—The Planet Jr. Double Wheel Hoe, Cultivator, Rake and Plow.

This tool is the best for all who raise onions or garden vegetables on an extensive scale. It does the work of six to ten men with ordinary hand hoes. It can be used to straddle the row, or between rows, as desired. It has all the attachments shown in cut. The Rakes level the ground for planting, gather stones and trash, cultivates, covers seeds, etc. The Hoes cut loose and clean, killing everything they meet, leaving the ground level. The Cultivator teeth mellow the soil deep or shallow, and are useful for marking out. The Plows lay out deep furrows and cover them, hill up or plow away, as desired. The leaf guards allow cultivation of large plants, such as beets, carrots, parsnips, beans and peas. No vegetable grower can afford to be without it. It will do the work of six men with ordinary hand hoes. Weight, 35 lbs...........Price, complete, $6.00

A233—Plain Double Wheel Hoe.—To meet a demand from some sections, the "Planet Jr." Double Wheel Hoe will be offered with only one pair of hoes and without leaf guards, as the "Planet Jr." Plain Double Wheel Hoe. Weight, 21 lbs...........Price, $3.50

A246—QUEEN OF THE GARDEN HAND CULTIVATOR.

Weight, 28 lbs.   Height of wheel, 23½ in.

Price...........$4.00

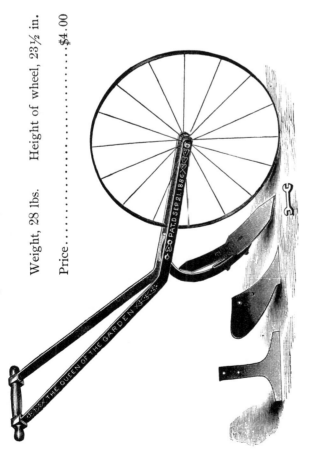

A247—THE DEXTER CULTIVATOR.

Single or double handle.

Weight, 17 lbs.   Height of wheel, 18 in.

Price...........$2.50

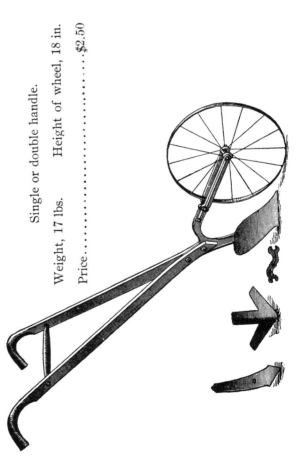

THE CELEBRATED MATTHEWS GARDEN TOOLS.
A239—THE MATTHEWS NEW UNIVERSAL MODEL GARDEN TOOL.

The Matthews Drill is designed to be used in field or garden. When in operation, it opens the furrow, drops the seed accurately at the desired depth, covers it and lightly rolls the earth down over it, and at the same time marks the next row, all of which is done with mechanical precision, by simply propelling the drill forward. In this way it sows, with an evenness and rapidity impossible for the most skillful hand to do, all the different varieties of Beet, Carrot, Onion, Turnip, Parsnip, Sage, Spinach, Sorghum, Peas, Beans, Broom Corn, Fodder Corn, etc.

The agitator stirs the seed in the hopper thoroughly by a positive motion, which insures continuous and uniform delivery, and the bottom of the hopper is made sufficiently dishing to sow the smallest quantity of seed. When desired, the movement of the agitator can be checked, and the drill may then be propelled forward or backward without dropping seed. There is al o an ingenious device by which the seed can be cut off while turning at the ends of rows, thus saving quite a percentage of seed. A simple contrivance accurately gauges the uniform deposit of the seed to any required depth thus avoiding the risk of planting at irregular depths, or so deep in places as to destroy the seed. The improved markers are made adjustable for the purpose of marking the rows at any desired distance apart, and they mark them distinctly whether the ground is even or uneven.

It is also provided with an indicator having the names of different seeds thereon To adjust the drill for planting different kinds of seeds, it is only necessary to turn the indicator around until the name of the seed to be planted comes to the indicator-pin at the top. This ingenious invention is a great improvement upon any other method in use, and is infinitely more convenient and reliable. The drill is simple in principle, and is constructed of the b st material and in the best style and finish, and, rightly used, will last many years and do a vast amount of service without requiring any repairs, as there are no parts subject to unusual wear. Price, boxed..each, $6.20

A240—MATTHEWS' COMBINED DRILL, HOE, RAKE, CULTIVATOR AND PLOW WITH SINGLE WHEEL.

This desirable combination of the Garden Drill already described with the new Universal Cultivating Attachments is one which cannot fail of recommending itself to all. The Attachments are 1 pair Hoes, 5 Cultivator Teeth, 1 pair Rakes and 1 pair Plows. The seeding and covering apparatus can be readily removed. It is not necessary to remove the seed box, as the teeth can be inserted in the frame with the box attached. Price, boxed..each, $7.50

A242—The Planet Jr. Combined Horse Hoe and Cultivator.

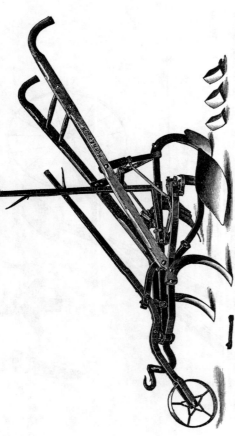

This well known and popular tool is acknowledged to be the best in the market. It has the *lever wheel*, a necessity for particular work; the patent *lever expander*, changing the width instantly; and *handle adjustment* for height; and also sidewise for use in grapes, pole beans, blackberries, nursery stock, etc., and when covering. It also has the patent parallel frame, interchangeable and adjustable *hollow steel* standards reversible blades, sleeved bolt holes, etc.; as a *horse hoe* with side teeth reversed it cuts close to the row without injury to the roots, leaving but little and in many cases nothing for the hand hoe. As a *cultivator*, the reversible teeth cut off and turn over the weeds in the most perfect style. As a *hiller*, the side teeth work in combination with the rear shovel to perfection.

Weight, 65 lbs. Price with both expanding and wheel lever................$7.20

A244—THE "FIRE FLY" GARDEN PLOW.

The tool is intended for those who have but small gardens and a moderate amount of time to spend in them. It enables them to raise vegetables for their family or for the market, with a minimum expenditure of labor and time, the latter being often the most important item to many who would be glad to grow their own vegetables if they could do it in their spare moments. The mouldboard is tempered and polished steel. The depth may be changed as desired very quickly. The low price brings it within the reach of all.

Weight 12 lbs. Price...$2.30

A143—The Wolverine Jr. Riding Cultivator.

The draft hitch is adjustable for hard or mellow ground, also to overcome the leading of the plows. The entire frame is of steel, neat, compact and light. Outside teeth with their section of frame can quickly be removed, reducing the Cultivator to six teeth. Steel points on the Cultivator are 2½ inches wide and 10 inches long, and are reversible, with break pin attachments.

Shields are furnished to protect small plants at first cultivation and turning plows for hilling when desired. All complete as here shown, with ninth shovel, or fallow tooth attachment.

Weight.........375 lbs.   Price.........$26.00

A153—Stationary Ratchet Handle Cultivator.
WITH AUTOMATIC BEAM SPRING.

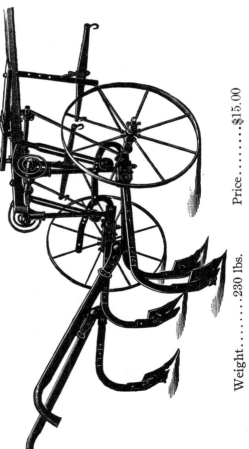

Weight......230 lbs.   Price......$15.00

See description of Beam Spring on next page. By loosening one nut the handles can be set at any position.

A154—Six-Shovel Power Handle Advance Cultivator.
AUTOMATIC BEAM SPRING.

Weight......230 lbs.   Price......$15.50

See description on next page. This Cultivator is the same as A 155, with the addition of two more teeth.

A16—Prairie Breakers, Slip or Bar Share, King of The Prairie.

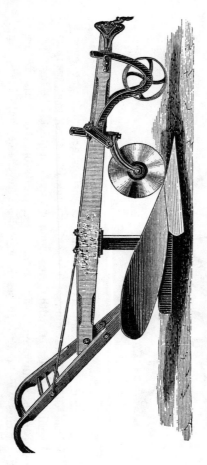

Right Hand Only.

Our King of the Prairie is the most popular Breaking Plow in use, and the easiest running. With long tapering mouldboard and flat share, it turns a flat, smooth furrow, with very light draft. The superiority of the material and most excellent turning qualities of our Breakers have placed them in the front rank of Breaking Plows. They are furnished with gauge wheel (and unless otherwise ordered) with our universal caster rolling coulter, three-horse adjustable clevis, and each Plow has an extra share fitted to it before leaving the shops, and a wrought frog. The clevis admits of an accurate adjustment for depth and land, and dispenses with gauge wheel clamps. Mouldboard solid cast steel. Share, cast steel, mild temper.

Inch.	Weight.	List.	Our Price.	Slip Shares.	Bar Shares.
12	130 lbs.	$20.50	$10.00	$2.00	$3.00
14	148 "	22.50	11.00	2.50	3.50
16	155 "	25.50	13.00	3.00	4.00

A88—M. W. & Co.'s Full Chilled Plows.

In all sections where Chilled Plows are in use farmers will readily recognize the above Chilled Plow, and appreciate the prices we quote for both the Plow and Repairs. Chilled Plows are used in clay, gravel and sandy loam. The shares do not cut as wide as the furrow turned, a part of the furrow slice being broken off at the bottom, instead of all being cut off with a wide share, as is the case with all steel plows on prairie soils. Two cast shares furnished with each of these Plows.

	Width at Share.	Width at Mouldboard.	Turns Furrow.	Depth of Furrow.	Weight.	Price.
A Right	7 inches	11 inches	8 inches	4¼ inches	50 lbs.	$3.60
B "	8 "	12 "	10 "	5 "	65 "	4.80
10 "	8 "	12 "	11 "	5½ "	70 "	5.25
13 "	9 "	13 "	11 "	6 "	80 "	6.00
19 Right or Left	10 "	15 "	12 "	6¼ "	100 "	6.25
20 " "	11 "	18 "	14 "	7 "	112 "	6.50
E1 " "	11 "	18 "	14 "	7 "	125 "	6.75
40 " "	12 "	20 "	16 "	9 "	130 "	6.75

Jointer, $1.50; Lead Wheel, 50 cents; Lead Wheel Standard, 50 cents. Both Wheel and Standard, $1.00 All cast Plows have shares narrower than the mouldboards. Only steel Plows have shares intended to cut as wide as the furrow turned. We always ship Right Hand Plows unless Left Hand is ordered.

A89—PRICE FOR REPAIRS FOR CHILLED PLOWS.
Warranted to fit any of the Chilled Plows we send out.

	Standard.	Mould-board.	Landside.	Shares, Plain.	Jointer Points.	Jointer Mouldboards
A Right	$0.95	$0.95	$0.30	$0.20		
B 10	1.13	1.32	.40	.21		
13	1.50	1.50	.45	.24		
19 Right or Left	1.60	1.68	.53	.25	15 cents.	40 cents.
20 " "	1.65	1.88	.55	.26		
E1 " "	1.68	2.10	.56	.27		
" "	1.88	2.25	.57	.28		

A48—STEEL BEAM SANDY LAND PLOW.

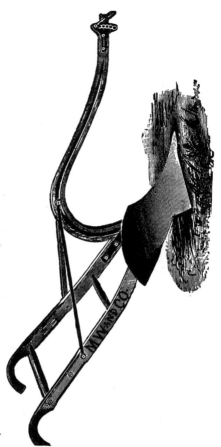

A50—WOOD BEAM SANDY LAND PLOW.

These plows are designed for the sandy lands of Texas and Arkansas, and are made very strong to stand hard and rough usage. They have been successfully used for several years past and give excellent satisfaction.

We make these Plows both wood and steel beam as shown. They have extra hardened mouldboard, crucible steel share and steel bar landside. They are made in the following styles and have one extra share:

	Wood Beam. List Price.	Our Price.	Steel Beam. List Price.	Our Price.	Average Weight.	Shares.
7-inch	$9.00	$5.00	$11.00	$6.00	73 lbs.	5 lbs. $ .90
8 "	10.00	5.50	12.00	6.50	80 "	6 " 1.00
9 "	12.00	6.50	14.00	7.50	90 "	7 " 1.20
10 "	14.00	7.50	16.00	8.50	95 "	8 " 1.40
12 "	16.00	8.50	18.00	9.50	110 "	9 " 2.00
14 "	18.00	9.50	20.00	10.50	120 "	10 " 2.25

A35—Grub Breaker and Township Plow.

As the cut shows, this Plow is compactly and strongly built. It has iron handles, and heavy iron plates on top and bottom of beam and iron shoe on side of beam to drag it on in turning. The cutter is solid cast steel and either end can be turned down and used. It is built with double standard, and the timber is *extra selected*. It is just what every township and road district needs for working roads. It is also an excellent Plow for breaking among grubs and clearing new land. Furnished with one share only. Size 12 inches, with iron handles. Price....................................$11.90 Weight 140 lbs.

A36—Brush Plow, with Cutter and Iron Strap Under Beam.

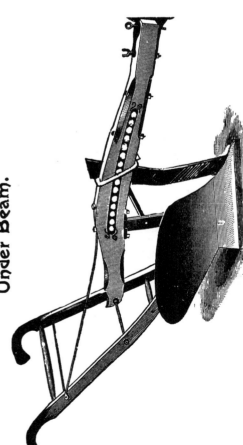

It is intended for new and brushy land where there are stumps and roots. It is very strong, and does its work perfectly in all kinds of land. It is also very generally used as a road plow, and as such gives uniform satisfaction.

	List Price.	Our Price.		Shares.
11-inch Brush, weight 78 lbs	$15.00	$7.75	8 lbs.	$1.90
12-inch Brush, weight 80 lbs	16.00	8.12	9 "	2.00
13-inch Brush, weight 85 lbs	17.75	9.00	10 "	2.25
14-inch Brush, weight 125 lbs	18.50	9.75	11 "	2.50

This plow is furnished with cutter as shown in cut without extra charge. This is a Quincy cutter, which extends upward and is bolted to beam, and can be reversed when one end is dull.

A46—THE YANKEE SWIVEL PLOW.

This Plow is designed especially as a side hill plow, but will work equally well on level land, turning the furrow all one way, and leaving no dead furrow. It is fitted with either Standard or Rolling Coulter, and with Chilled Iron Mouldboard. A foot lever secures the plow in position. When the plow is to be reversed the driver merely touches the lever with his foot and the turning of the team reverses the plow from right to left as desired.

The clevis is governed by the rod projecting backward to the handles, so that the driver has perfect control of the width of cut.

This is the latest improvement, and best reversible plow there is made. Made for one, two or three horses. Two shares go with each Plow.

	Plain	With Wheel	Wheel and Cutter	Wheel & roll'g Coulter	Extra Shares
No. 0 is a one-horse plow, so arranged that the horse walks in the furrow, and turns a furrow 4 to 6 inches deep and 9 to 11 inches wide. Weight, full-rigged, 85 pounds	$ 6.35	$ 7.12	$ 8.00	$	$0.35
No. 1 is a light two-horse plow, or when ordered, can be arranged for a large one-horse, and turns a furrow 5 to 6 inches deep and 10 to 12 inches wide. Weight, full-rigged, 105 pounds	8.60	9.00	10.72		.50
No. 2 is a two-horse plow, and turns a furrow 6 to 7 inches deep and 12 to 14 inches wide. Weight, full-rigged, 117 pounds	9.50	10.25	11.50	12 25	.60
No. 3 is a two-horse plow, and turns a furrow 6 to 8 inches deep and from 13 to 16 inches wide. Weight, full-rigged, 123 pounds	10.00	10.75	12.00	12.75	.65
No. 4 is a three-horse plow, and turns a furrow 7 to 9 inches deep and from 15 to 18 inches wide. Weight, full-rigged, 172 pounds. By using with this size plow, three-hors. Equalizer, three horses can be placed abreast	11.50	12 25	13.50	14.25	.7

A24—ROD BREAKER—STEEL BEAM.

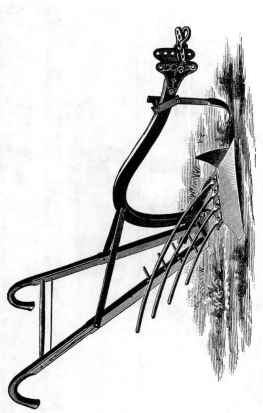

Right Hand Only.

	Weight.	List Price.	Our Price.	Shares.
12-in., with Steel Landside, Fin Cutter and Extra Share,	75 lbs.	$14.00	$5.75	$1.50
14-in., " " " " " "	78 "	15.00	6.00	2 00
16-in., " " " " " "	80 "	16.00	6.25	2.50

This plow is not made as a substitute for the Mouldboard Breaker in general use, nor is there claimed for it any great superiority, but it is a thoroughly well made plow, and in the light prairie sod of some localities it completely fills the place of the Mouldboard Breaker, and has the advantage of being much cheaper. It is very light and easily handled, and of very easy draft.

In place of the Mouldboard it has four wrought iron rods or fingers, so shaped that their turning quality is practically the same as the Mouldboard of A 16.

Shares and Landside are of the best quality of high natural tempered cast steel, and provided with fin cutter.

Shoe gauge is easily adjustable.

A52—Cotton King, Black Land, Texas Plow, Wood Beam.

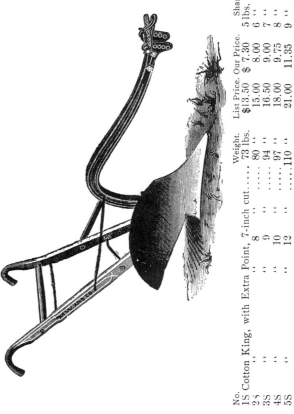

This is our Cotton King Plow with new shape boards, adopted after a thorough investigation of the requirements of a plow for the blacklands of Texas and the South.

No.		Weight.	List Price.	Our Price.	Shares.	
1W	Cotton King, with Extra Point, 7-inch cut	54 lbs.	$10.50	$5.75	5 lbs.	$ .90
2W	" "	62 "	12.00	6.50	6 "	1.00
3W	" "	79 "	13.50	7.25	7 "	1.20
4W	" "	82 "	15.00	8.25	8 "	1.40
5W	" "	100 "	18.00	10.00	9 "	2.00

A54—Cotton King, Steel Beam.

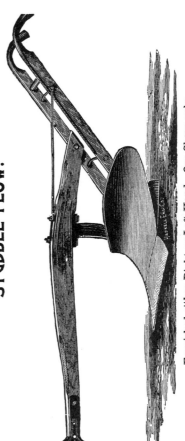

No.		Weight.	List Price.	Our Price.	Shares.	
1S	Cotton King, with Extra Point, 7-inch cut	73 lbs.	$13.50	$ 7.30	5 lbs.	$ .90
2S	" "	80 "	15.00	8.00	6 "	1.00
3S	" "	94 "	16.50	9.00	7 "	1.20
4S	" "	97 "	18.00	9.75	8 "	1.40
5S	" "	110 "	21.00	11.35	9 "	2.00

A43—NATIONAL STEEL BEAM TURF AND STUBBLE PLOW.

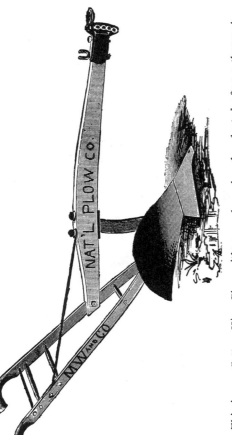

This Plow has long tapering mouldboard with easy turn, and will plow tame sod "just right." Also a splendid plow for old ground; and for clay soil it cannot be excelled. They are extra hardened in share, landside and mouldboard—finely polished and painted. Both right and left hand in following styles:

	Weight.	List Price.	Our Price.	Shares.	
12-inch cut, double shin	90 lbs.	$17.25	$ 9.35	9 lbs.	$2.00
14-inch cut, double shin	100 lbs.	20.00	11.00	11 lbs.	2.25
16-inch cut, double shin	110 lbs.	22.50	12.00	12 lbs.	2.75

A44—NATIONAL WOOD BEAM TURF AND STUBBLE PLOW.

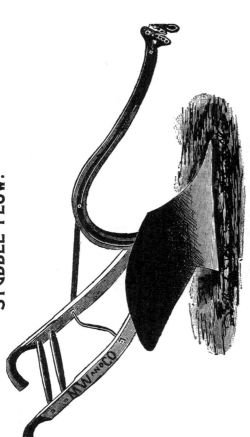

Furnished either Right or Left Hand. One Share only.

These Plows are sometimes called general purpose plows, and are especially suitable for farmers who rotate their crops, and require the same plow for both sod and stubble. Has double board, and tempered as hard as any plow we sell. Plows are very light draft and do fine work.

	Weight.	List Price.	Our Price.	Shares.	
12-inch cut, double shin	80 lbs.	$15.75	$ 8.50	9 lbs.	$2.00
14-inch cut, double shin	90 lbs.	18.25	10.00	11 lbs.	2.25
16-inch cut, double shin	100 lbs.	20.50	11.00	12 lbs.	2.75

A104—5-TOOTH CULTIVATOR.

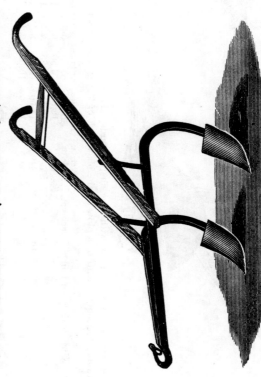

Wood Frame, Weight 60 lbs..........List Price, $6.00. Our Price, $3.00. Extra teeth, each 30 cents. Standard brace, 35 cents.

Diamond teeth are 5 inches wide and 7 inches long. Cultivator spreads 31 inches. Teeth are reversible. By loosening nut on the back, either edge can be brought in use, or teeth can be removed for sharpening.

A106—Steel Beam Double Shovel Plow.

Iron, Weight 40 lbs..........List Price, $4.00. Our Price, $2.50. Shovels..........50 cents.

Teeth are 6 inches wide and 11 inches long, hardened steel. Width of Cultivator, 20 inches.

A76—CONTRACTOR'S PLOW.

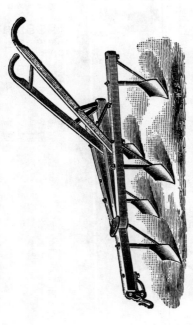

This Plow is built especially for contract work, very strong and durable, and is used by leading contractors throughout the United States and Canada. Designed for from two to six horses.
Weight.....155 lbs. Price.....$10.00 Points, each.....$0.75

A82—DITCHING PLOW.

A very strong and rapid digger; will loosen dirt in bottom of ditch that is four feet deep, doing the work of several men. It never fails to give satisfaction. Does not dig the ditch, only does the work of a pick. The loosened dirt is thrown out with a shovel. Price..........$10.00
Weight..........120 lbs.

A3—OUR NATIONAL PLOW BOY.

Right Hand Only.

			Shares.	
12-in., with old land bottom,	$25.50;	with turf and stubble bottom,	$26.50	9 lbs. $2.00
14 " " " "	26.50;	" " " " " "	27.50	11 " 2.25
16 " " " "	27.50;	" " " " " "	28.50	12 " 2.75

Prices are with pole, whiffletrees and yoke; average weight 420 pounds. Breaker bottoms furnished as an extra attachment, including two breaker shares, 12-inch, $10.00; 14-inch, $11.00; 16-inch, $12.00.

In the construction of this Plow we dispense with the heavy frame work as cumbersome and unnecessary. It has a large land wheel and has a slanting rear wheel, which is necessary to take pressure off the land-side. The levers are convenient, and the Plow being perfectly balanced, it raises easily and high, the coil spring rendering efficient aid. Spring allows the wheel to pass over stones or obstacles without jarring or causing the Plow to come out of the ground.

This Plow will draw lighter with a man on the seat than a walking plow of the same dimension and it rides easily.

A7—HAPGOOD'S THREE-WHEEL SULKY PLOW.

Its reputation alone usually proves sufficient to sell this implement, but to those not acquainted with its merits we would say that it is warranted to turn a square corner better than any other sulky made. It is light draft, simple in construction, very easy to handle and made of the best material. Made with steel wheels, with movable boxes and has large rolling coulter. In turning at the ends, or moving about, the doubletrees raise high off the ground, and the brake keeps the Plow from running on the horses' heels. Weight, 500 pounds.

Price 14-inch steel share, tongueless.............$28.30
  "   16   "   "   "   " 29.00

We can furnish cast shares on the above when so ordered.

A8—THE "OLD RELIABLE" RIDING LISTER.

With new Triple Cut Off, Spring Lift, Adjustable Spring Chain Tightener and Automatic Throw-out. Is the only Riding Lister made that will double list. Being high under the beam and having a leveling lever, it will run where others cannot work. Raising the Lister out of the ground with the lever throws the dropping apparatus out of gear, so that corn is not scattered in turning or moving. It is furnished with three-horse eveners that enable the operator to put the outside horse in the last row and thus gauge the rows all the same width. It drops 12, 14, 16 or 20 inches apart. The Subsoiler can be adjusted to any required depth. Covering blades throw up a ridge in the center of the row, leaving a drain on either side, preventing the washing out of the corn, or being covered up by a washing rain; in fact, it has every necessary point or feature to make it a Perfect Lister.

Price, 14-inch Sulky Lister, complete, with covering shovels..........$35.50
  "   16   "   "   "   "   "   " 37.00

A92—O. K. PLOW SULKY ATTACHMENT.

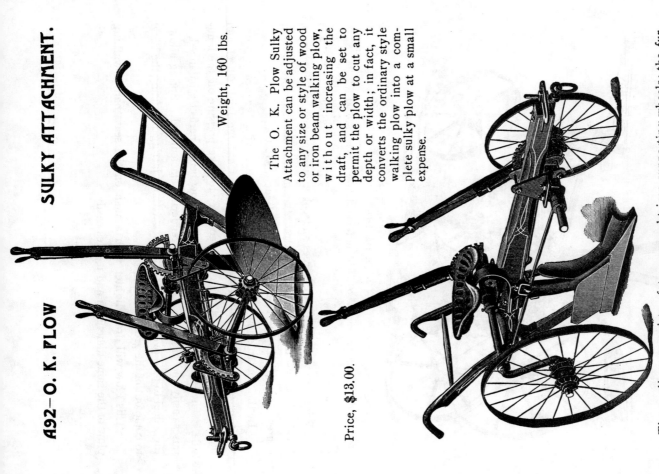

Weight, 160 lbs.

Price, $13.00.

The O. K. Plow Sulky Attachment can be adjusted to any size or style of wood or iron beam walking plow, without increasing the draft, and can be set to permit the plow to cut any depth or width; in fact, it converts the ordinary style walking plow into a complete sulky plow at a small expense.

The appliance consists of two wrought-iron supporting wheels; the furrow wheel is 24 inches in diameter, and the land wheel is 30 inches in diameter; substantially constructed of the lightest and best material, accurately fitted to the axle and seat frame; and the whole being so arranged as to be strongly clamped to the plow beam. There is nothing about the attachment to get out of order; it requires no mechanical ability to put it on or to take it off. It will enable the plow to operate satisfactorily in any kind of soil and under all conditions. Attachment alone; no plow furnished.

A140—"National" Adjustable Arch Cultivator.

This Cultivator has springs, which makes it very easy to handle. The hitch is direct, so that each horse does his share, and being adjustable on the arch, can be set so shovels run as close to plants as desired. As soon as draft of team is applied the drag-bars rise up, and when team quits pulling and gangs are hung up, the drag-bars bear on ground.

IT MAKES A PERFECT CULTIVATOR

and can be used with spring tooth gangs, as shown in cut, or with plain steel or wood beam gangs or eagle claws, as desired.

Weight 175 lbs.

Shields are furnished, and steel wheels 30 inches high.

Price, with Spring Tooth Gangs...................$13.50
"   "   Steel Beam   " 12.50
"   "   Wood Beam   " 12.50
"   "   Eagle Claw   " 13.00

A142—M. W. & Co.'s New Riding Disc Cultivator.

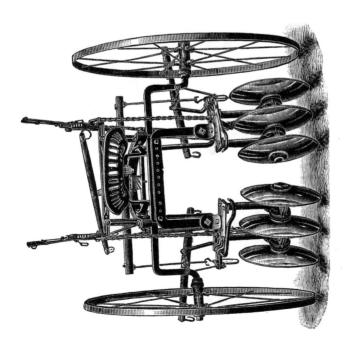

| Weight 440 lbs | 14-in. Disc | Price $25.00 |
| " 450 " | 16-in. " | " 27.00 |

This cultivator possesses all the advantages of our other riding cultivators, and has in addition the cultivating discs, which can be set at any angle and any distance apart, and can be reversed so as to turn the ground to or from the corn. Has low hitch. Each horse pulls his side of the cultivator. No swinging of tongue.

The wheels and gangs work parallel and are thrown side-wise by simply bearing down on a foot-pedal, leaving both hands free to manage the team. The dodge is instantaneous. This makes plowing of the crookedest rows very easy, overcoming a great fault with all other discs. We hitch direct to the front end of the beam with a clevis, the same as to a stubble plow, and are enabled to gauge the depth just the same as on a walking plow.

A100—Miller's Bean Harvester.

This Harvester has gained an enviable reputation in the bean raising districts wherever used. It surpasses and has a greater sale than any other machine of its kind. It has no equal in the completeness of its work.

By means of the rods on the flexible, rolling dividers, the vines are gathered and brought together into a windrow at the rear of the machine, free from roots and dirt.

The adjustment of the machine is simple, and may be operated, as to depth, easily from the seat by means of the levers.

The drive wheels are provided with ribs in the centre of the rim, which prevents the machine from slipping sideways, either on level ground or on hillsides.

The guards in advance of the drive wheels remove all loose stones from their paths, which would otherwise raise the machine and be a hindrance in the performance of its work.

Price..$30.00
Weight 300 lbs.

A101—National Clevis Spring Cultivator.

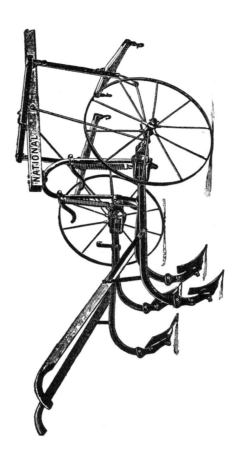

This Cultivator is as well made, nicely finished, and of as good material as any on the market. Fully warranted in every respect. Has steel beams and steel wheels.

Price, with four shovels, or eagle claws..........$13.00
" " Spring Teeth Gang like A140.......... 14.00
With parallel beams, extra.......................... .50

A116—LITTLE GIANT 14-TOOTH HARROW.

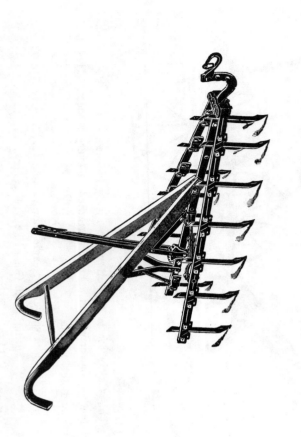

This is the most complete and perfect tool of the kind in the market, combining, as it does, a Field Cultivator, Garden Harrow and Pulverizer. Teeth are so arranged that one end is the Cultivator and Pulverizer, while the other end is the Harrow. By a very simple device the slant of the teeth can be changed so that the tool can be made a perfect Smoothing Harrow. It has fourteen $5/8 \times 7/8$ diamond teeth, drawn to a cutting edge on one end. Cultivator can be spread from 12 to 31 inches.

Price, with lever as shown $4.50
Leading wheel extra .... .50
Extra teeth without Clips .... .08

Singletrees.

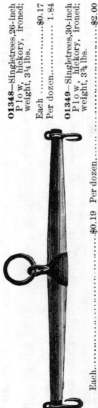

Each .... $0.19 Per dozen $1.84
O1350—Singletrees, 36-inch, wagon, hickory; ironed with ferrules and hooks; weight, 6 lbs.
Each .... $0.32 Per dozen $3.25

O1348—Singletrees, 26-inch Plow, hickory, ironed; weight, 3¼ lbs.
Each .... $0.17 Per dozen $1.84

O1349—Singletrees, 30-inch Plow, hickory, ironed; weight, 3¼ lbs.
Each .... $0.19 Per dozen $2.00

O1351—Wagon Neck Yokes.

A strong, heavily ironed Neck Yoke, 38 inches long; weight, 6¾ lbs. Each $0.40
Per dozen $4.50

O1352—A set consisting of evener, singletrees, neck yoke and clevises. All ironed complete and painted one coat. Weight, 30 lbs. $1.50

Perfection Plow Whiffletrees.

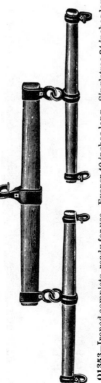

O1353—Ironed complete, ready for use. Evener 40 inches long. Singletree 34 inches long. Suitable for general farm work. Adjustable clips. Are made of best seasoned hickory, thoroughly oiled, and the best quality malleable clips. No holes are bored in the wood to fasten clips, thus preserving full strength of the wood. Price, per set $1.50

Perfection Wagon Doubletrees.

O1354—Ironed complete, with stay chain clips and plates on both sides of evener; woods oiled. Evener 48 inches long. Singletrees 36 inches long. Has adjustable clips. Suitable for wagons, threshers, engines, water tanks, etc. Quality guaranteed. Price, per set $1.75

Combined Two and Three-Horse Eveners.

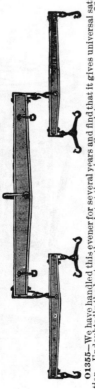

O1355—We have handled this evener for several years and find that it gives universal satisfaction. Undoubtedly the best on the market at the price. Complete $2.00

O1360—The Dandy Plow Doubletree

Hickory Evener, 38 inches long, with patent clips. A very fine Doubletree for ordinary work.
Ironed, complete $0.90

OUR CHANNEL STEEL HARROW.

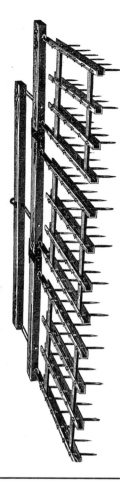

Made in two sections, 48 square teeth; three sections, 72 square teeth; four sections, 96 square teeth. Is simple and durable. It has no bolts or nuts to work loose. The teeth reverse automatically by hitching to the other side of frame, and it does perfect work in all kinds of ground. Has hinged draw bar and same steel rail as on our Lever Harrow.

	Cultivates.	List Price.	Our Price.
A 190, 48 teeth, weight 130 lbs	6 feet	$13.00	$ 6.00
A 191, 72 teeth, weight 200 lbs	9 feet	19.00	9.00
A 192, 96 teeth, weight 250 lbs	12 feet	25.00	12.00

WOOD FRAME VIBRATING HARROW.

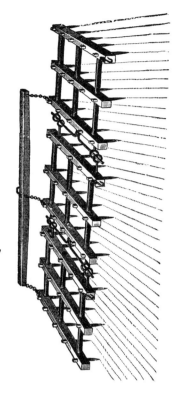

Oak beams and spools, steel teeth, one-half inch square. Size of timbers 1⅜ x 2¾, bolted through and through; turned wooden spools. We furnish extra teeth to either of the above Harrows at 3½ cents each. Teeth are ½-inch steel, 9 inches long; no other size furnished. Sections have four bars, 20 teeth to a section.

	Weight.	Price.
A 194, 1 section, 20 teeth, cuts 3½ feet	50 lbs.	$ 2.50
A 195, 2 sections, 40 teeth, cuts 7 feet	100 lbs.	5.00
A 196, 3 sections, 60 teeth, cuts 10½ feet	150 lbs.	7.50
A 197, 4 sections, 80 teeth, cuts 14 feet	200 lbs.	10.00

A201—Our Spring Tooth Steel Frame Harrow Without Levers.

Observe that the frame is protected by the lower curve of the teeth. Harrow can be regulated to any required depth of cut.

No. 21, steel frame, 16 tooth.	Weight 120 lbs	each, $11.50
No. 22, " " 18 "	" 140 "	" 12.50

A201½—Our Lever Spring Tooth Steel Frame Harrow.

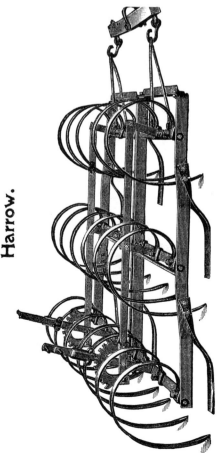

The teeth are adjusted by shoes (not by revolving or changing the pitch) which gives the teeth the proper pitch at all times and regulates them to the required depth, at the same time making a lighter draught by keeping the frame away from the ground. The shoes will raise the harrow to all heights and take the points of the teeth four inches from the ground for dumping or transportation.

No. 40, steel frame, 16 tooth, lever attachment. Weight 160 lbs	each, $13.00
No. 41, " " 18 " " " 180 "	" 14.00
No. 42, " " 20 " " " 200 "	" 15.00

A102—DISC HARROW WITH BROADCAST SEEDING ATTACHMENT.

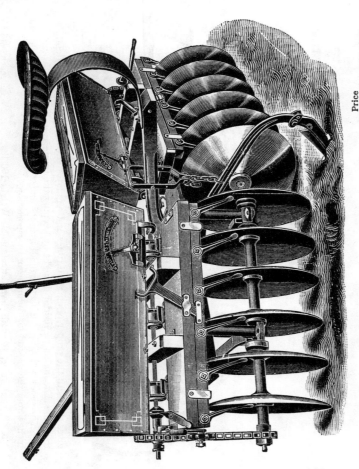

	Weight.		Price Broadcast Sower.	With Drill Attachment.
With twelve 16-in. Discs,	500 lbs., with four-horse eveners		$33.00	
" sixteen 16 "	600 " " " "		40.00	$40.00
" twelve 20 "	600 " " " "		35.00	
" sixteen 20 "	700 " " " "		43.00	50.00

The above cut represents our Improved Disc Harrow, with new Seeder attachment, which is a great improvement. We claim for it the following advantages over all others:

1st. It is driven from both sections, and the drive chain is always in line.
2d. It is provided with a spring chain tightener that enables you to take the chain off or put it on without removing the wheels, and prevents it coming off.
3d. Either side can be thrown out of gear when desired to finish a land or sow a narrow strip.
4th. It is easy to throw out of gear.
5th. The driver sits where he can see at all times if the seeders are discharging properly.
6th. It has a perfect index, showing the quantity sown to the acre.
7th. It is nicely and substantially put up in every way.
8th. We have a center shovel that covers the grain sown in the middle, which no other has, and they not only leave a ditch, which we do not, but they leave grain in the center that is not covered.
9th. It can be used without a shovel if desired.

A174—IMPROVED IRON FRAME DISK HARROW.

The frame is neatly and strongly made of wrought-iron bars and connects to the disk gangs by castings which form a universal joint. The pivot being back of the axles causes the gangs to move toward the center as they are given the proper angle by moving the inside ends back, thus bringing them together, which leaves no ridge uncut in the center, or at least reduces it to a minimum.

There is a set of balls at each end of each of the four boxings, both balls and boxings in which they work being thoroughly chilled, rendering them so hard that they will not cut out, and do not require any oiling whatever, thus making a saving of both time and trouble of applying it.

Twelve	16-inch Discs,	cut 6½ feet,	weight	350 lbs.
Fourteen	16 "	" 7½ "	"	400 "	$22.00
Sixteen	16 "	" 8½ "	"	425 "	25.00
Twelve	20 "	" 6½ "	"	400 "	22.00
Fourteen	20 "	" 7½ "	"	450 "	25.00
Sixteen	20 "	" 8½ "	"	500 "	28.00

First line price: $20.00

Above prices include 3-horse Evener.

A176—THE "BUDLONG" OR "LA DOW" DISK HARROW.

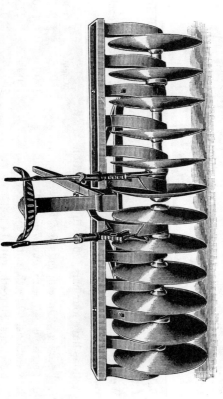

The accompanying cut shows our STEEL DISK HARROW, made under the celebrated "LaDow" patents, and is the only Disk made that the boxing will not wear out with the side pressure. The two inside ends of the sections butt together and support each other, and they will last longer than the others. All others wear out in a short time. The boxing is oiled through the top of the standards, putting the oil just where it is wanted, and where the dirt and sand can't interfere. We build them with steel weight boxes, which is necessary, as there are times when a Disk is no good without them. Many others have none, and are surely not perfect machines. They draw from the center of the axle, and are warranted to have no side draft or weight on the horses' necks. Warranted to give satisfaction. Arranged to use three or four horses, and we will furnish either style, as ordered, at the prices quoted below:

No. 1—Twelve	16-inch Disks,	cuts 6 feet,	Price	$21.00
No. 2—Fourteen	16 "	" 7 "	"	22.50
No. 3—Sixteen	16 "	" 8 "	"	24.75
No. 4—Twelve	20 "	" 6 "	"	22.75
No. 5—Fourteen	20 "	" 7 "	"	25.60
No. 6—Sixteen	20 "	" 8 "	"	27.00

A256—SHOE PRESS DRILL.

Price, 8-Hoe.....$53.00   12-Hoe.....$76.00   16-Hoe.....$103.00
Grass seed attachment, $2.50 extra.   Shipping weight about 600 lbs.

This Drill has steel wheels and steel frame in one piece, being bent out of one continuous piece of steel pipe.

There is not a joint in the entire frame, hence no weakness in any part of it.

It has the best fit feed, and has the best pressure on the shoe.

Pressure can be put on the shoe alone, or pressure can be put on the shoe and press wheel jointly.

The press wheel can be taken off entirely and chain or other covering device fastened on the shoe in its stead.

The pressure on each shoe is separate and distinct, and Drill will adapt itself to uneven surfaces, covering more perfectly than any machine that uses the shoes in gangs.

The grass seed box is datachable, so that when the farmer is done seeding it can be removed and emptied of the remaining seed. Made in this way, it also prevents water from getting into the seed by running down the side of the grain box. Underneath the grass seed hopper is placed an adjustable deflecting board, by means of which the grass seed may be scattered either in front or behind the hoes at will.

The whiffletrees are attached below the frame in such a way that the draft tends to take the weight off the horses' neck,

A202—NATIONAL CORN PLANTER.

Price, with Hand Hill Dropper and Drill..................$31.00
" Hand Dropper, Drill and Check Rower..................35.00

1st—It is a combined Check, Drill and Hand Drop Planter.
2d—It is a complete Hand Drop Planter.
3d—It is a complete Drill Drop Planter.
4th—It is a complete Check Rower Planter.
5th—It can be changed into any one of these three combinations immediately.
6th—The number of kernels to be planted is regulated by the number of holes in the plate.
7th—It is of the most simple construction.
8th—It is made of steel, with the exception of the tongue, and is therefore the most durable Planter made. Weight 175 lbs.

INSTRUCTIONS FOR OPERATING.

In putting the plates in the hopper, always see that the same size holes are in both plates and the same number of holes in both plates, and that the plate is dropped into the hopper so that the opening in the plate is exactly over the opening in the shoe. In other words, put both plates in the planter alike.

Eight-hole plates will drop four to five kernels or drill 11 inches apart.
Seven-hole plates drop three to four kernels or drill 13 inches apart.
Six-hole plates drop three kernels or drill 15 inches apart.
Five-hole plates drop two to three kernels or drill 18 inches apart.
Four-hole plates drop two kernels or drill 22 inches apart.

We can furnish plates for dropping other kinds of seed at an extra price. Bear in mind it is necessary in selecting your seed corn to have the butt and point grains shelled off if you expect accurate planting. Then try your Planter until you get a plate with holes of proper size to drop your corn, and if no plate be found with holes of proper size, ream the plate to fit the corn, always reaming from the bottom side.

A245—The "Victor" Potato Coverer and Cultivator.

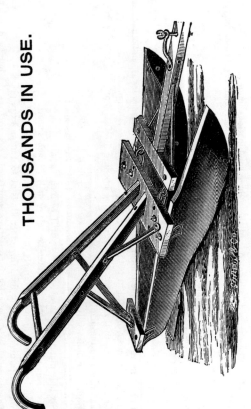

THOUSANDS IN USE.

WHAT IT WILL DO.

A team of horses and one hand (it is light work for a boy) will cover *from six to ten acres a day*, in either check, rows or drills, or dropped on the surface, the amount depending on the speed of the team, the width of planting, and the previous tillage. The work will be done better than a gang of men would do it with hoes. The operation of the implement is to form slight ridges over the rows, and these ridges are left smooth, evenly finished and compact. The soil is left in the best possible shape to insure quick and certain germination of seed.

By the time the young potato plant breaks through the ground, the farmer finds that a vast crop of weeds is likewise coming up. Here the use of the "Victor" will save a large amount of future trouble and cost at a trifling expense. It not only saves labor, but by cultivating thoroughly it insures a large subsequent yield. The right thing to do at this important period, is to cover the potatoes a *second time* lightly, and thus every weed between the hills and also in the hills is destroyed. This second covering, which is done as rapidly as the first, seems to impart great vigor to the potato plants, and they break through the ground strong and healthy and grow rapidly. The use of cultivators will thereafter keep the crop free from weeds, without resorting to hand-hoeing. When the time arrives for hilling the potatoes the "Victor" is again called in use, and it does this work in a remarkably superior manner. The ground between the rows should first be put in fine order by cultivators, and then the Coverer, or as it now might be termed, the Hiller is used. It accomplishes the work in a somewhat peculiar and very effectual way, rolling up the soil to the hill and *tucking it under the* plants, destroying all small weeds, and surpassing ordinary hand labor.

In this simple and low-priced tool, we offer the Farmer an effectual implement for three distinct and important operations necessary to grow a crop of potatoes, namely, for Covering, Cultivating and Hilling.
Weight, 70 pounds. Price, complete.................$8.00

A97—CLIPPER POTATO DIGGER.

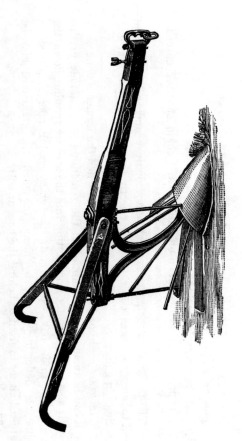

Weight, 97 lbs.   Price, $7.00.

These diggers are put up with three rods on each side, with extra plates attached for two rods, which are sometimes used where the soil is heavy; these rods must be raised at the heel as shown in cut With our cast standard we bolt the shovel and beam rod securely in position, from which it cannot yield, as in wood standard and shoe. As shown in cut, the projecting part of standard under the beam is firmly held in position, and the point of shovel is always straight with the beam.

The depth of the digger is regulated by the rod from heel of shoe through end of beam. Where the potato vines are heavy use a sulky rake lengthwise of drills, which will tear and pull out the vines, then follow with the digger in the same direction.

A204— KEYSTONE CORN DRILL.

This is a light and compact machine, all the working pieces being made of steel and malleable iron.

The gearing is protected; all changes can be made rapidly and safely.

The operator can always see the corn dropping.

There are no loose gearings that can get out of order.

It can be adjusted to sow one corn grain 10, 12½, 15 or 18 inches apart, or two grains at 20, 25, 30 or 35 inches apart.

It will sow Indian corn, broom corn, lentils or beans.

Weight, 100 lbs. Price........................$10.00
Extra for fertilizer attachment, $4.00.

A212—The Chautauqua Corn, Bean and Seed Planter.

Planter Entering the Ground.
JAWS CLOSED.

Planter as it Should Leave the Ground.
JAWS OPEN.

The movement of planting with this machine is very similar to that of walking with a cane, and with the spring lightly adjusted, is but little more labor. The spring may be adjusted lightly or stiffly by the nuts holding it down. Keep the end of the blade in the center by the relative tension of these two nuts. Weight, 4½ lbs. Price....$1.25

CORN PLANTERS.

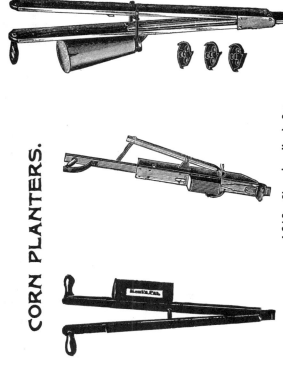

A214—Triumph Corn Planter.

Kent's Patent Triumph Hand Corn Planter, known and used everywhere; can be used on sod or plowed ground Weight, 4 lbs.

Price, each $0.75

A216—Champion Hand Corn Planter.

Has pumpkin seed attachment. Glass slide shows just how many kernels are ready to drop in next hill. Every miss hill can be avoided; no blank hill with this Planter. Weight, 4½ lbs. Price.........$1.50

A217--Eagle or Eclipse Rotary Drop Hand Corn Planter.

Four discs furnished of various sizes of holes, moved by both a push and a pull pawl; action positive and sure; plants accurately. Weight, 4½ lbs. Price..........$1.25

A218—Cotton Seed Planter & Fertilizer Distributor.

All metal parts in this Planter are steel in place of iron. Implement is materially improved. Too well known through the Southern States to need description. Price.................$2.25

A103—SUCCESS CORN HARVESTER.

The Success Corn Harvester.

The cutting wings are pivoted to the center of the Harvester and are readily and quickly thrown in under the platform with the levers by the operators while in motion, standing on the platform, and thus close them quickly to pass gallows hills, or obstructions, or to prevent an accident to man or horse. This is a safety and advantage possessed by no other, and is worth more than the price of the machine. The levers are convenient to the operators and do not require any stooping down or back breaking to handle them, and either operator can throw the steel wings in or out at his will and pleasure. The convenient location of the levers and their easy movement enables the operator to cut the last hill next to the shock and throw in the wing to pass the gallows hills of shocks without stopping. This places the harvester past the shock and leaves the shock free from obstruction and convenient for the operator to deposit his armful of stalks. This is another advantage possessed by the Daisy, making less stops and enabling the operators to cut from twenty-five to fifty more shocks per day than anyother. Average weight 400 lbs.

No. 3 with levers, no wheel..$11.00
No. 4 with both levers and wheel, like cut........................ 12.00
No. 5, lever, wheel and shafts.................................... 16.50

A166—Riding Corn and Cotton Stalk Cutter.

	List Price.	Our Price.
Single Row, 2-Horse, shipping weight 350 lbs.	$40.00	$22.00
Double Row, 3-Horse, shipping weight 650 lbs.	65.00	40.00

Cut shows the new patent Single Row Corn and Cotton Stalk Cutter. Double row has two revolving discs, machine just double the width of the single row Cutter; has double pole to use three horses.

The plan is different from all others, and instead of cutting by the weight of the Cutter Head and machine, the power comes entirely from springs. The combination of lever and springs enables the operator to apply any force to the knives that may be necessary to do the work, and when raising from the ground to turn, the same spring raises the Cutter Head, making it the easiest Cutter to handle in the world.

The seat and wheels are high, and the driver well up out of the dust.

The wheels running into low places have no effect on this cutter, as the frame never rides the Cutter Head, which has always plenty of room to work, and the knives are never crushed into the ground as with all other cutters, making them do poor work and overstraining the horses.

The steel wheels have movable boxes which are bored out and fitted up as perfectly as those on a buggy. The axles are also turned upon a lathe with a hub nut instead of a hole in the axle with a pin to keep the wheel on.

When offered cheap stalk cutters, it is well for the purchaser to consider if it will not pay better to buy machines made of good material, well finished, having wheels with adjustable hubs and perfect in every respect.

A254—The M. W. & Co.'s Force Feed Seeder.

The feed is the most vital part of a seeding machine, and a machine that has a poor feed can not do good work nor give satisfaction. We furnish a seeder with an absolute force feed, the efficiency of which is not changed when set to sow a large or a small quantity.

The fluted feed rolls render the use of an agitator unnecessary. Every seed cup sows the same quantity because the feed rolls are all substantially attached to the same shaft. Our feed has no set screws to get loose and render them trappy and unreliable.

The seed spouts are long, and bring the grain near the ground. The spouts are hinged to the frame so as to avoid accident.

Our improved spring hoes are so constructed as to instantly resume their proper position after passing an obstruction. We are aware that cheap seeders can be had at most any price; we offer ours to our customers with the assurance that it has no superior in point of efficiency, finish or construction. Weight about 500 lbs.

Price............10-Hoe, $38.00. 12-Hoe, $41.00. 14-Hoe, $43.00.

A255—11-FOOT BROADCAST SEEDER.

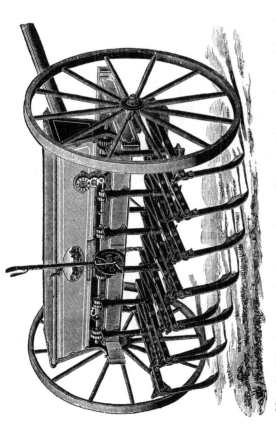

Weight 350 lbs. Price...$14.00

A262—The Granger Broadcast Hand Seed Sower.

Sows on an average of six (6) acres per hour, at a common walking gait. A person entirely unused to sowing by hand can use this machine with perfect success and do much better work than can possibly be done by hand, and as the seed is distributed so much more evenly there is a great saving in seed over hand sowing.

The bag and hopper will hold about 22 quarts—as much as a man would wish to carry.

Retail price............$5.00
Our price............. 3.25
Weight, 5½ lbs.
A263.—The Improved Cahoon Broadcast Hand Seeder. Our price $3.00

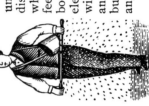

A264—THE LITTLE GIANT HAND SEEDER.

This is somewhat simpler and cheaper made than the Granger, but works on the same general plan. It will sow with ease to the operator 60 acres of grain or grass seed per day. Has a shake feed, and sows perfectly accurate.

Weight, 4 lbs. Price $2.25

A266—The Little Giant Broadcast Seed Sower.

This is a very convenient and substantial seeder and undoubtedly the easiest operated of any seeder made. It will distribute flax and clover seed thirty-six feet to the round; wheat, fifty feet to the round; timothy seed, twenty-seven feet to the round; oats, thirty-six feet to the round. The bow is three feet long and imparts a very high motion to the eleven-inch distributing wheel. It will sow any kind of seed without recleaning, including orchard and blue grass seed, and from the smallest quantity up to two and one-half bushels of oats to the acre. Gauge set by thumb screw for any quantity, with convenient and perfect cut-off.

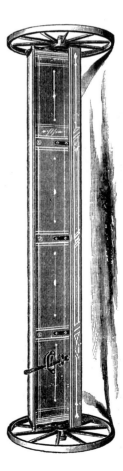

Weight, 9 lbs. Price.. ....$1.50

A252—KNAPSACK SPRAY PUMPS.

For Orchard, Vineyard or General Use.

A Convenient Outfit with which from 4 to 5 Acres of Vines can be Covered in One Day.

The Myers Knapsack Spray Pump.

This cut represents the Myers Knapsack Spray Pump. The tank holds five gallons, is fitted with lid and strainer, which can be removed.

It is a Brass Bucket Spray Pump, with air chamber, ball valves, solid plunger and agitator.

We wish to call attention to the fact that this is the only Knapsack outfit on the market **with an agitator.**

It is so arranged that no water can drip on the operator. The pump can be removed easily. The operator can set the knapsack on the ground and work pump with handle. It can be carried by hooking snap in staple on the opposite side, provided for that purpose, making a neat handle, as shown by dotted lines.

The handle lever can be shifted from right to left shoulder, at will.

Price, $5.50.

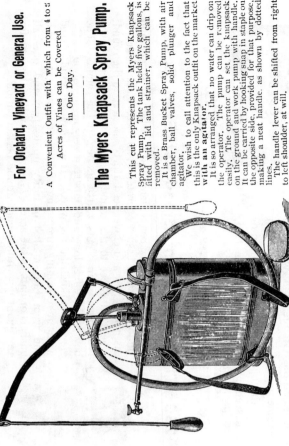

A First-Class Outfit at a Reasonable Price.

A253—M. W. & CO.'S 5-HOE WHEAT DRILL.

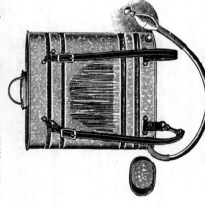

Weight, 140 lbs. Price, $14.00. With Fertilizer Attachment, $22.00.

It is simple, light, strong, durable, easily adjusted, and sows splendidly.

This popular Drill for sowing wheat, oats and barley between rows of standing corn, has been extensively and favorably known for years.

The Feed, being the most important part of any Drill, is of course to be mentioned first, as without a perfect feed all other points are of little account. Our Drill has an adjustable force feed, by which the quantity can be varied without change of gear wheels. It is a positive force feed, and nothing can be more regular and accurate in its work.

The feed can be shut off at any time by a lever in the rear, throwing Drill out of gear.

The depth can be regulated by the gauge wheels at rear of frame, which form also a truck complete, except barrel, consisting of pump, 5 feet of 1/2-inch 3-ply hose, Vermorel or Bordeaux spray nozzle, 3 feet of suction pipe, strainer and agitator..................... 6.30

A250—Myers Knapsack Spray Pump.

Pressure made with hand on rubber bulbs; can reach small trees as well as low plants. Furnished with one pipe and spray nozzle.
Price, one hand, one nozzle...........$3.00

Extra Hose and Spray Nozzle, used on Knapsack Pump, No. A 250. Price.........$1.00

A250½

A251—BARREL SPRAY PUMP.

The usefulness of this Pump is not limited to the spraying purposes only as described, but it is constructed with base adapted to be fastened on sinks and platforms, so it can be used for a house hose pump in cisterns, shallow wells, or at any place where a neat, powerful force pump is required. When used for house purposes, it is furnished with a discharge spout, elevated to the proper height.

WHITEWASHING.—The Myers Spray Pumps have proven themselves a decided success for whitewashing, and are worth more than their cost for this purpose in whitewashing factories, cellars, fences, chicken houses, etc., and can be utilized to great advantage in places which could not be covered by the brush or by the old method at all. The high pressure put on the fluid causes it to penetrate all the unevenness and crevices in the surface, covering same uniformly.

We do not furnish barrels; procure a good, sound kerosene barrel and mount pump as here shown. Our Utility Cart, A725, can be used in this connection to advantage.

Price, pump only, with strainer and agitator......................... $4.85
" complete, except barrel, consisting of pump, 5 feet of 1/2-inch 3-ply hose, Vermorel or Bordeaux spray nozzle, 3 feet of suction pipe, strainer and agitator........ 6.30

A274—Improved Pacific Broadcast Seed Sower.

For sowing wheat, oats, barley, grass seed and any other cleaned seed; also plaster, salt, ashes, commercial fertilizers, etc. Sows one-third wider cast than other machines.

Weight, 135 lbs.   Price, $9.50

It is a complete machine, not an attachment, and we warrant it to sow one-third wider than other leading machines now in use, and more evenly. It is constructed so that no seed strikes the tail-board, falling behind the wagon and leaving the grain to grow in streaks, thus wasting much seed. This machine is placed in wagon and can be set at any angle by raising or lowering one end of frame. The Seeder can be stopped or started when in operation; can sow on one side when desired, and is instantly regulated for any kind of grain or seed.

The gearing and all wearing parts of Pacific are very much stronger than any other, making it far more durable—the velocity of distributing wheel being much faster than any other, throws the seed 33 per cent. farther; this we guarantee. The seed hopper is much larger than others. Furnished with wood or iron hoppers.

The Cast: The Pacific has a double set of slides, one for graduating the amount to be sown, the other to stop or cut off the flow entirely. The graduating slides are operated separately, so a greater or less quantity may be sown on either side when used in strong wind, or the sowing may be all on one side when finishing a field. The regulated device is perfect.

One Hundred Acres of Wheat Can be Sown in Ten Hours.

Quantity Sown per Acre: This seeder has a gauge to show the quantity sown per acre, but the amount sown is governed altogether by the speed of horses. Some horses walk twice as fast as others, consequently the distributing wheel moves twice as rapidly, and sowing very much wider, and would require the slides to be opened in proportion to speed of team. All farmers know from appearance of grain on the ground how wide to open slides.

A276—Niagara Broadcast Seeder.

A Positive Force Feed End Gate Seeder.

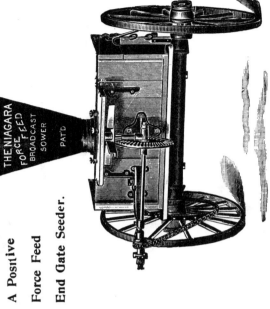

Weight, 125 lbs.   Price, $8.50

The "Niagara" combines all the latest improvements in sowers. The machine is attached to an end gate that fits any wagon. Attach sprocket to left hind wheel, put on link chain and go ahead. Save grain and seed by sowing it even. Time is money. Early seeding is what tells. Don't spend your time with the old style seeder when four times the amount of seed can be evenly distributed with the Niagara. This machine throws the seed evenly and fully two acres to every half mile traveled by the team. The Niagara has all the good qualities of other broadcast sowers.

A277—Our "Western" End Gate Broadcast Seeder.

Weight, 125 lbs.   Price, $7.50

This Seeder is manufactured for the trade in immense numbers, each jobber having it branded some fancy name selected by himself for his particular trade. It is known by a dozen different names, and is in use in every State under one name or another. We will call it "Our Nameless," rather than add one more new name to this standard and well-known implement. Warranted a perfect Broadcast End Gate Seeder.

A1004—ROUND TANKS.

				Weight.	
2 ft. Stave,	5 ft. Bottom,	200	8 Bbls.		$ 7.50
2 "	2¼ "	250	11 "		8.12
2¼ "	5¼ "	300	11 "		8.75
2¼ "	5¼ "	325	14 "		9.37
3 "	5¼ "	350	16 "		10.65
4 "	5¼ "	400	22 "		13.50
2¼ "	6 "	350	14 "		10.00
2¼ "	7 "	450	22 "		13.00
2¼ "	8 "	475	30 "		14.25

Covers for round tanks, 5 ft.............................. 2.00
"   "   "   over 5 ft........................ 2.50

Considering 50 gallons to barrel,

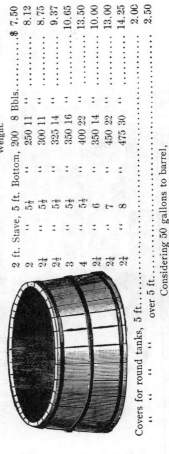

A1006—SQUARE TANKS.

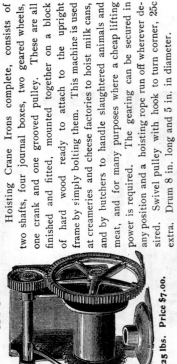

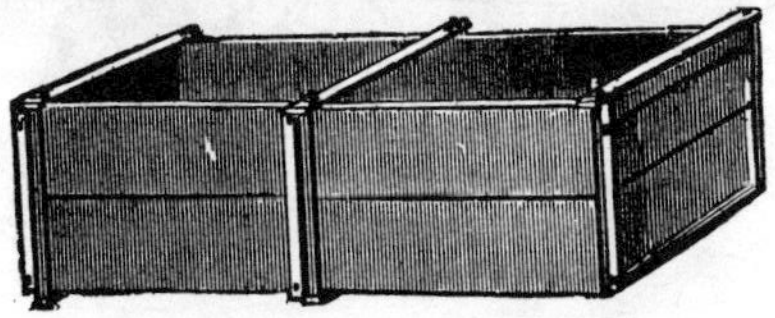

	Barrels.	Weight.	
11 feet long, 34 inches wide, 20 inches high	11	450	$10.15
12 " " 34 " " 24 " "	13	500	12.25
13 " " 40 " " 24 " "	18	575	14.35
15 " " 52 " " 24 " "	20	650	17.10
16 " " 52 " " 20 " "	22	700	19.00

A1012—HOISTING CRANES.

Hoisting Crane Irons complete, consists of two shafts, four journal boxes, two geared wheels, one crank and one grooved pulley. These are all finished and fitted, mounted together on a block of hard wood ready to attach to the upright frame by simply bolting them. This machine is used at creameries and cheese factories to hoist milk cans, and by butchers to handle slaughtered animals and meat, and for many purposes where a cheap lifting power is required. The gearing can be secured in any position and a hoisting rope run off wherever desired. Swivel pulley with hook to turn corner, 25c extra. Drum 8 in. long and 5 in. in diameter.

Weight 125 lbs. Price $7.00.

CAST-IRON PIG TROUGH.

Indestructible and Everlasting.

In each trough there are eight separate compartments, eight large hogs can eat at one time. Different kinds of feed can be fed at the same time, the bowls preventing their mixing. Hogs cannot upset them and waste their feed, nor is there any leaking, but great economy and satisfaction in using these troughs.

A1013—Weight, 120 lbs. Height, 24 inches. Width, 32 inches. Price.............$4.90

A 248—Western Potato Bug Sprinkler.

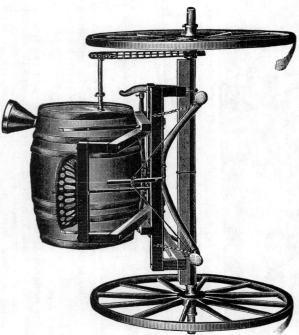

It applies the poison upon two rows at once, and will cover twelve to fifteen acres per day. This is a machine that has given general satisfaction wherever it has been used. It is one of the cheapest, simplest and most perfect Potato Bug Sprayers ever invented.

Any boy that can drive a horse can perform, with more ease, the same amount of work that heretofore required twelve men in performing, in the same amount of time.

You have complete control over the quantity of water you wish to throw, opening or closing the valve with a lever. The spraying heads can be adjusted to any position desired, up or down, in or out, all of which can be done while the machine is in motion. The wheels run on tubular axle and can be shifted to any width of rows from two to three feet. The agitator is driven by chain belting from the wheel. The chain wheel on the agitator can be shifted in or out. This will keep the poison from settling to the bottom.

Weight, 450 lbs. Price............................ $25.00

A 249—Gray's Patent Paris Green Sprinkler.

This is a *Gravity* Sprinkler; force comes from elevation of tank. Useful for potatoes, currant bushes, rose bushes and low shrubbery.

It is painted inside and outside, is provided with capsules (roses), by means of which the stuff is economically sprinkled over the leaves in form of fine spray. It has an apparatus by which the stuff and the water are kept mixed constantly. The quantity to be used is regulated by pressure of the little fingers on the rubber hose at the connection of rubber and tin pipe.

Weight, 8½ lbs. Price............................ $3.00

A726—HAND CART.

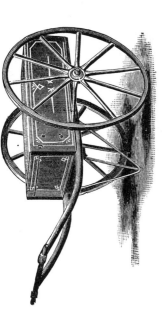

36-inch wheels; box, 24 x 36, 10 inches deep. Weight, 85 lbs. Removable end board, bent handles, iron foot rest, iron hubs, well painted and striped. A first class job. Very useful about barn, stable or garden. Price............$5.00

A732—WOOD FRAME BARREL CART.

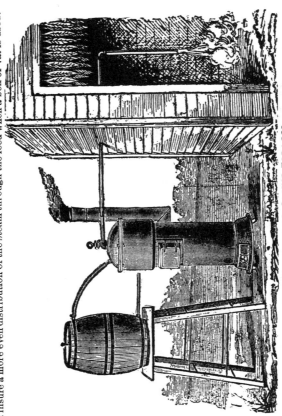

We furnish the trucks only, to be attached to any kerosene or molasses barrel. Useful for gardening and feeding purposes. Large flanges to bolt to any barrel. Two of these carts can often be used to advantage; one kept near the kitchen to be filled with skimmed milk or slops, while the other is being emptied at the piggery. A patent device to attach to different barrels costs more and is not as convenient as a set of wheels and handles to each barrel. On account of heavy weight and short twists to which wheels are subjected, wood wheels prove stronger than iron or steel wheels for this purpose. Weight, 65 lbs. Can furnish all steel wheels and steel handles if preferred. Price......$3.25. Weight 65 lbs.

A733—STEEL FRAME BARREL CART.

Made entirely of steel, no wood being used in its construction. Furnished with irons for attaching to barrel. Price each, $3.30, Weight 49 lbs.

A735—STEEL HAND CART WHEELS.

To parties who wish to make their own hand carts we offer a nice light steel wheel. Tire, 1½-inch half oval, 38 inches high, to ft 1⅛-inch axle. These are very nice wheels for hand carts.

A706—"Triumph" Farm Boiler or Steam Generator

Our STEAM GENERATOR can be used for steaming feed in a barrel or a hogshead as represented in the above illustration, or a tight box or vat with a cover will answer the purpose better where it is advisable to cook a quantity of feed at a time. A false bottom perforated with holes will insure a more even distribution of the steam through the feed when a box or vat is used.

FOR ORDERING OR CASING TOBACCO.

The above cut represents the STEAM GENERATOR ordering tobacco. A large number of the generators have been sold in the South for this purpose. By actual test, size No. 2 will order a tobacco barn 16 ft. square, and containing several hundred pounds of tobacco, in about an hour.

DANVILLE, VA., Sept. 6, 1889.

The Triumph Steam Generator ordered through you last spring has proven entirely satisfactory. I have used it only for ordering dry lots of tobacco to prevent its breaking when being prized and the result is equal to a natural season.
C. C. DULA, (Tobacco Broker.)

Prices Delivered F. O. B. in Chicago.

Size.	Horse Power.	Diameter.	Height.	Size of Fire Box.	Shipping Weight	Price Complete.
1 G	¾	15 in.	48 in.	10x24 in.	350 lbs.	$37.00
2 G	1	17 in.	50 in.	12x24 in.	450 lbs.	48.00
3 G	2	22 in.	56 in.	16x26 in.	800 lbs.	60.00
4 G	2½	22 in.	63 in.	16x33 in.	900 lbs.	70.00

A302—THE PERFECTION HAY TEDDER.

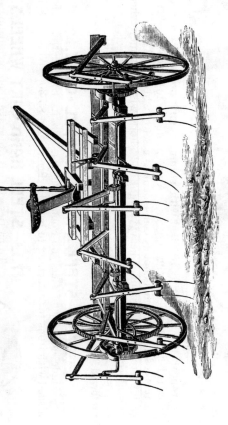

This Tedder does effective work, throwing the hay sufficiently high, and leaving it on the ground open and loose so as to permit a free circulation of air. Operated by a boy the Perfection Tedder will do the work of ten men, and cure the hay more effectively and in much quicker time than can be done by hand.

A new method of attaching the outside kicker arm to the frame instead of the extended axle, thus removing all possibility of loosening the main shaft and causing bad breakages. There has been introduced a very useful feature in the shape of a device which allows the main frame to react or move upward behind when the forks strike the ground or other hard substance. This you readily see will save the forks from breakage.

The method of throwing in and out of gear is the simplest in use, and can not possibly get out of order.

The Perfection Hay Tedder has two levers—one for raising and lowering the frame, and the other for throwing in and out of gear. Both can be readily used by the driver from his seat without dismounting. Combined pole and shafts.

We should have a week's notice before shipment. Do not put off ordering a Tedder until the day you want to use it.

Tedder for one horse, can also be used with two; six forks, weighs 450 lbs. Price............$23.00

8 Forks. 500 lbs. Price............$28.00

A304—THE PERFECTION 8-FORK TEDDER

FOR TWO HORSES.

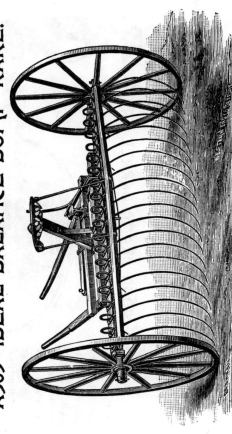

Combined pole and shafts, can be used with one horse same as No. A302, but stronger and wider. Weight, 500 lbs. Price............$31.00

A305—IDEAL BALANCE DUMP RAKE.

The "Ideal" Balance Dump Rake, made with twenty coil steel teeth of the highest grade of steel carefully tempered. It has wrought iron wheel spindles, and the axle is stiffened by a truss rod which prevents any sagging. In operative qualities it will meet the requirements of the farmer, being a perfect lock lever, which can be operated with great ease, while the teeth can not rise except when required to dump The workmanship is first-class; made with wooden wheels only.
Weight, 200 lbs. Twenty teeth, 4¾ inches apart, price............$11.55

A307—THE SELF-DUMP HAY RAKE.

The Most Ingenious and Effective Dumping Mechanism Ever Used on a Rake.

This rake is very similar to our No. 310 All Steel Rake, only it is made of wood where steel is used on No. 310.

On these rakes we use a new patent tooth with a sled runner point that cannot scratch the ground or take up grass roots. The point of the tooth runs about one inch above the base of the tooth. The teeth are made of the best steel and are thoroughly tested before leaving the factory. Teeth about 4¾ inches apart.

Price, wood wheels, wood axle and shafts, 20 teeth...$17.00
" " " " 26 " ... 20.00
Steel wheels instead of wood, extra............. 1.00

A310—ALL STEEL SELF-DUMP RAKE.

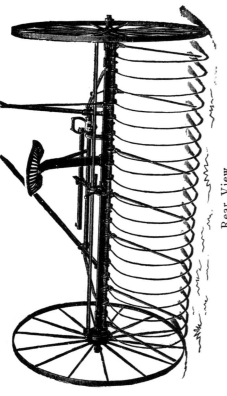

Rear View.

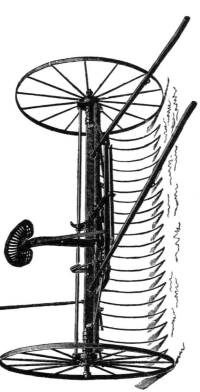

Front View.

This rake is made entirely of steel, no woodwork whatever being used in its construction. Just the implement for hot or dry climate and all sections where the woodwork is subject to deterioration on account of a severe climate. These rakes will last a lifetime.

Price, with 20 teeth......................$19.50
" " 26 " 22.00

A311—ALL STEEL HAND DUMP RAKE.

This rake is made on the same principle as No. 310, but must be dumped by hand. Has combined Pole and Shafts.

Price, with 20 teeth......................$16.75
" " 26 " 19.75

SHOWING CARRIER AT WORK ON BARN WHERE HAY IS TAKEN IN FROM THE END.

When building a new barn it would be well to frame it in a "Ridge Pole," and let the same extend 4 to 6 feet beyond the roof to support the track. An iron brace will answer the same purpose. Make the window large and suspend the door by means of pulleys and balance weights so as to slide up and down.

A365—SNATCH PULLEY BLOCK.

This device shortens the travel of the horse without reducing the elevating power. Tie a knot in the rope so that when the hay is going up the horse has a direct pull; but when the carrier is unlocked and the hay begins to move along the track the rope doubles around the snatch pulley attached to the whiffletree; you will see that the horse travels only one-half the distance the hay is carried, a great saving; the rope can be thrown off of the snatch pulley and fork be instantly returned to the load without waiting for the return of the horse. Thus: When the horse pulls direct on the rope, the rope takes up just as fast as the horse travels; but when the rope doubles around the pulley, the rope takes up twice as fast as the horse walks away. Weight, 5 lbs.
Price..$0.60

A324—DOUBLE HARPOON HAY FORK.

No farmer who has ever used the Horse Hay Fork for unloading hay, either in the barn or on the stack, will ever go back to pitching by hand. Those who neglect to adopt it are wasting money and time every year. The best double Harpoon Fork we sell for 63 cents, and ropes, carriers, hooks, pulleys, etc., at prices below competition. Save your backs and let the horses do the work. The whole outfit can be secured for a few dollars—cost depending on size of barn.

Double Harpoon Hay Fork, all iron and steel and very durable. The best fork in the market for many kinds of hay; known and in use everywhere. Weight, 18 lbs.
Price..$0.63

Length of tines, 24 inches.

A326—Long tine double Harpoon Fork, the same as above, only heavier, and has tines 32 inches from cross bar. Weight, 22 lbs. Price, each........$1.30

FOUR-TINE GRAPPLE HORSE HAY FORK.

Width 17 inches. Opens to 4 feet 6 inches.

SIX-TINE GRAPPLE HORSE HAY FORK.

Width 17 inches. Opens to 4 feet 6 inches.

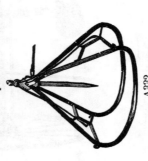

A332.

A334.

A332—Four-tine Grapple Horse Hay Fork. Weight, 40 lbs. Each......$4.25
A334—Six-tine Grapple Horse Hay Fork. Weight, 55 lbs. Each......5.25

A336—14-TINED FORK FOR HANDLING MANURE.

Every other tine to this manure fork can be removed, when it will be exactly like A334 and adapted to handling hay. Set your stacking outfit over your manure pile and load manure by horse power. One boy and horse and one man can do the work of ten men with hand forks. Width, 23 inches. Opens to 4 feet 6 inches.
Price..$0.25

A337—Spear points to above three forks. Weight, 3 lbs. Price........$0.50

Same as A334, but wider and more tines. Weight, 87 lbs.

A342—HARPOON HAY FORKS.

The Nellis Improved Single Harpoon Hay Fork. Weight, 7 lbs. Price............$1.25

ROPE HAY SLING OUTFIT.

For Unloading Hay and Grain Without Fork.

A344—Sling Pulleys.

To be used with our new Leader or Milwaukee Carriers. Steel Track, Wood Track or Cable Track, with our standard wagon sling. Weight, 10 lbs. State which carrier the pulley is to match. Price for pulleys only as shown in cut............$2.00

These pulleys at $2.00, and three slings shown below $6.75, make the outfit in place of fork cost $8.75.

SLING PULLEY.

A346—STANDARD WAGON SLING.

Used in Connection with Sling Pulley above, in place of Horse and Hay Fork.

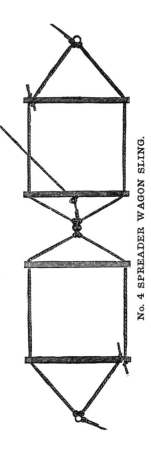

NO. 4 SPREADER WAGON SLING.

Weight 18 lbs. Price, each............$2.25

Two of these slings remove a load in two parts, one sling being placed on bottom of wagon, and the other in the middle of the load. Many use three slings and hoist the load in three divisions.

Slings can be made lighter by using only two slats or spreaders.

SHOWING CARRIER AT WORK ON 4x4 TRACK WHEN WAGON IS ON BARN FLOOR.

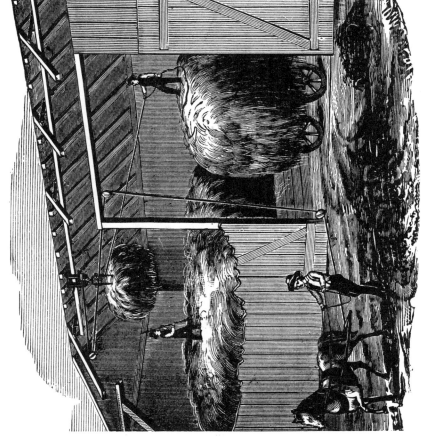

Use hanging hooks on every rafter over the floor and every other rafter over the bays. Carrier works either way by merely changing end pulley to the other end of the barn, carrying rope with it, or you can thread the rope through the carrier as shown on page 86, attaching the horse at the other end. Make your own estimate from this cut as to the number of hanging hooks and length of rope wanted. When horse is at door sill rope should be long enough to allow fork to touch the floor.

This cut shows hanging hooks supported by wooden cleats. Our rafter irons cost 5 cents each, and brings the carrier nearer to the peak of the barn, making more room for large forkfuls of hay.

A316—ADVANCE HAY LOADER.

We offer this Hay Loader with the distinct assurance that its possession will place its owner in the best possible position for taking care of his hay crop in the most economical, safe and expeditious manner that modern invention has yet pointed out.

The value of a Hay Loader is now conceded by every intelligent farmer. Such a general interest has been aroused in this labor saving machine that every one who has enough meadow land to justify the outlay is looking for a loader.

The Hay Loader represented above is offered to our customers with the guarantee that it will, in every particular, do its work as well if not better than any one now in use. It is offered with the further guarantee that it is constructed of better material, is better made and more handsomely finished than any other now on the market. These are only statements of facts that can easily be verified, without any expense, by simply giving the loader a trial.

It is intended principally as a swath loader, but will work very satisfactory in windrows that are not too heavy. In fact, where the hay is light it is better to rake it in small windrows first.

An examination of the above cut will show the extreme simplicity of the loader. This simplicity so reduces the draft as to make it a perfectly practical machine in all kinds of hay and on all kinds of ground. The manner of attaching to the wagon is the simplest that has yet been devised, and it can be detached by the person on the load without first descending to the ground.

Weight, about 800 lbs. Price................................$50.00

A814—NEW M. W. & CO. STEAM OR BELT POWER PRESS.

This machine has given giving perfect satisfaction in every instance.

The gearing is double, very strong and simple. The crank gear uses three-quarters of a revolution in pushing the plunger in, and one-quarter in withdrawing the same (the plunger returning three times as fast as it goes in), giving long time to feed and an enormous power at the time required, enabling the operator to drop in the followers or dividing boards with perfect safety to himself and the press, without a moment's loss of time.

A five or six horse power engine being as large as required. It has no condensation box or other clap-trap devices; all that is required in the way of feeding is to push the hay or feed over the opening under the packer. Each press has an alarm gong, which rings at the time for placing the dividing board in the chamber between bales. These can easily be dropped in place without stopping the press.

This press is constructed of the very best material, finely finished in every respect, and having all conveniences necessary to do rapid work.

A bale every minute. Size of bale at end, 17x20 inches. It will put into a bale three feet long, 150 to 175 pounds, or a less quantity, to suit the party balnig.

Price, F. O. B. factory, $235 net, cash with order, no discount.

Weight, about 5,000 pounds.

A294—Montgomery Ward & Co.'s Wide Cut Mowers, with Bar-Carrying Spring.

Weight 675 lbs.

We make a sustaining spring to carry the weight of the cutter bar on the driving wheels, still leaving the cutters flexible to conform to the ground, decreasing both forward and side draft by transferring the weight, otherwise dragged sideways on the ground, to the driving wheels, and increasing the power of cut by increasing the weight.

The force of spring may be varied by the nut at its machine end.

A spring arranged to do less must be faulty, if not worthless. More than accomplished in these devices is not necessary. It seems impossible to gain like results more simply or efficiently.

The gain in use of bar-carrying springs is beyond belief until seen in the field. Then, the reason why a six-foot or seven-foot bar mower, with these fixtures, *exhibits equal cutting power for each inch in width* without increase of draft, or side draft, over the narrow mowers without them, is a surprise until we consider that excessive draft and side draft is the result of ground friction which is entirely avoided by this plan of construction.

The sickle eye is large; has no parts or bolts to give trouble in getting loose.

The oscillating joint of bar permits a self-adjusting tilt of guards to ride into and out of dead furrows, or if on top of a bog the guards will point down to cut clear its opposite side, making thus a *genuine floating bar.*

Brass filings are supplied for crank journals, and for wrist pin bearing, which may be easily renewed.

The driver's seat is behind the cut, out of danger, easily accessible, convenient to levers, and arranged to balance weight on wheels instead of horses' necks, and adjustable forward and backward to suit driver.

The general arrangement is for completeness in use, easy action for man and team, simplicity, neatness and endurance,—*a good, common-sense first-class mower.*

Price. 4 ft. 6-in.cut.................$40.00
" 5 ft. cut...................... 42.00
" 6 ft. cut...................... 45.00

A708—The Great Western Feed Steamer.

40 inches long. Weight, 350 lbs. Price, $28.50.

It is the very best thing made for heating water for slaughtering purposes; for making soap; for laundry and dairying purposes.

It will get up steam in ten to twenty minutes, and cook a barrel of potatoes, turnips or feed in from fifteen to twenty minutes.

WALTER A. WOOD TWINE BINDER.

With Single Apron Harvester and Bundle Carrier.

We ask the attention of thinking farmers to the *points gained* in the Walter A. Wood SINGLE APRON Harvester and Binder not enjoyed before:
Greater compactness and firmness.
Decreased weight.
Gain of about one-third in ease of draft.
Time saved by avoiding stoppages.
Increased cleanness of harvesting.
No grain shelled in elevating.
Grain brought to the binder in better shape and therefore better bound.

A763—THE NEW FAMOUS CORN SHELLER AND SEED CORN BUTTER AND TIPPER.

Weight, 12 lbs. Price.....$1.08

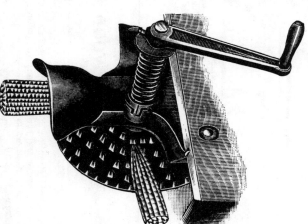

This is a most simple and effective arrangement for a hand Corn Sheller. There are only six pieces in the entire machine: a main frame, sheller disk, shaft, coil spring, tension ring and crank.

This Sheller separates the cob from the corn, discharges the corn into the box, on which it is mounted, and the cob outside. Corn to shell well in any machine must be dry and well cured. We have in this sheller a perfect tension that can be adjusted instantly. Will shell all sizes of ears Pop Corn to the largest southern dent.

THE SEED CORN BUTTER AND TIPPER ATTACHMENT.

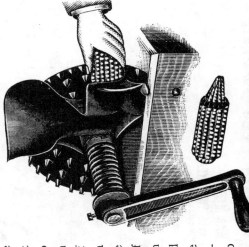

If accurate planting of corn, by hand or horse machine, is desired, it is necessary to remove the butt and tip kernels which are always uneven in size and imperfect, before the corn is shelled for seed. We have in this Sheller a perfect Butter and Tipper, for which no extra charge is made, and which is worth more than the price of the Sheller. It consists of a rest cast in the frame of the Sheller (see cut). The ears of corn, *butt* and *tip*, are placed in this rest, and the Sheller turned a few times backward, when the work is easily and quickly accomplished. Will butt and tip about fifteen ears per minute.

A296—THE BOSS SICKLE GRINDER.

Note the low price we are making on this implement.

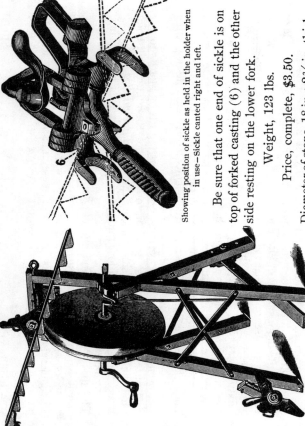

Showing position of sickle as held in the holder when in use—Sickle canted right and left.

Be sure that one end of sickle is on top of forked casting (6) and the other side resting on the lower fork.

Weight, 123 lbs.
Price, complete, $3.50.
Diameter of stone, 18 in.; 2¾ in. thick.

We are making a leader of this desirable implement, so that any one who has use for a good grindstone and owns also a mowing machine cannot afford to do without it. It costs but a trifle more than an ordinary grindstone and will answer for all purposes. We do not sell parts of this machine, Order the complete implement. The stone is accurately centered and hung true, and does perfect work.

A297—PERFECTION SICKLE GRINDER.

The Perfection we believe to be the simplest constructed, easiest understood, and one of the most eficient Emery Wheel Sickle Grinders manufactured. Weight, 18 lbs. Price, $3.00

A 298. The well-known Dutton Sickle Grinder, latest improved pattern. Weight, 16 lbs. Price, $6.00

184

A584—ORIOLE FARM GRIST MILL No. 2 1-2

Weight, 150 lbs. Can be run with one or two horse power. For grinding fine corn meal and graham flour for table use. It is made in the most durable way possible, runs easy, and has an automatic feed.

This mill will grind from five to eight bushels per hour. One set of burrs will grind 2,000 bushels of grain before they are worn out. They can be replaced for $1.25. Speed of mill is about 800 revolutions per minute. Size of pulley 8 inches; diameter of burrs 7 inches. An extra set of burrs furnished with each mill free of charge. Does not grind ear corn.

Height, 3 feet. Price..........$20.00
Extra burrs per pair..........1.25

A586—WILSON'S No. 3 FARM MILL.

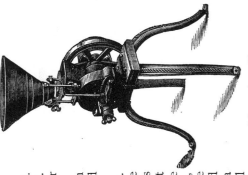

Power required, from two to four-horse. Regular sized pulley, 8 inches in diameter. We furnish a 15-inch pulley when desired. Speed of mill from 600 to 700 revolutions per minute.

This mill will grind green bones direct from butcher, dry bones, greasy bones, corn and cob and grain.

Capacity on bones from 600 to 1,000 lbs. per day. Capacity on grain from five to twelve bushels per hour. This mill contains two sets of burrs. The first breaks the bones to about the size of corn grains, after which they are then taken out and the second ones put in, which grinds them into meal. The shafts are made of steel and the burrs of chilled iron, and are put together in sections, so that when a part is worn out it can be duplicated at small cost. A special set of burrs for grinding grain goes with each mill. Weight about 300 lbs. Height, 3 feet 7 inches. Price..........$45.00
Burrs per set..........2.00

A562—$3.80 FAMILY GRIST MILL.

Weight 30 lbs. Price, each..........$3.80
Removable burrs..........75 cents per pair.*

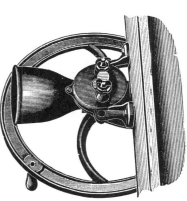

NEW PATTERN.   OLD PATTERN.

We have sold thousands of these mills and shipped them to every state in the Union. They are well adapted for grinding corn meal, hominy, split peas, cracked wheat, graham flour, fine table or butter salt. In fact, everything that is ground at a custom mill, except fine bolted family flour.

This mill will pay for itself in tolls you pay your miller. It will pay for itself because you will have samp, hominy, graham and fresh corn meal, when, without it, you would very often have to use wheat flour instead.

We have made great improvements in all of our hand Grinding Mills. In the old mills the burrs were cast solid on the main shaft and on the casing. In the new mills the grinding plates are removable and can be changed in a moment's time. They are of the hardest cast steel. Our hand mills will grind just as much *in proportion to area of grinding surface and speed used* as any large power mill, from ½ lb. to 1 lb. per minute being the average, according to kind of grain and fineness. We warrant these mills fully equal to any hand mill of same size made.

A564—Montgomery Ward & Co.'s Duplex Mill.

Without Stand. 20 inches high.

This is essentially the same mill as No. A562 above, but combined with our improved famous corn sheller.

Shells corn and grinds grain. Shelled corn or meal passes into receptacle below. The sheller is the most perfect implement of its kind; it does not scatter corn; cobs are thrown out to the rear of the machine and entirely separated from the shelled corn; shells and grinds with remarkable ease. B is the corn sheller hopper; A is the grain grinding hopper.

Price each, $4.80. Weight, 55 lbs.

Extra burrs, per pair, 75 cents.

A623—KELLY DUPLEX GRINDING MILL.

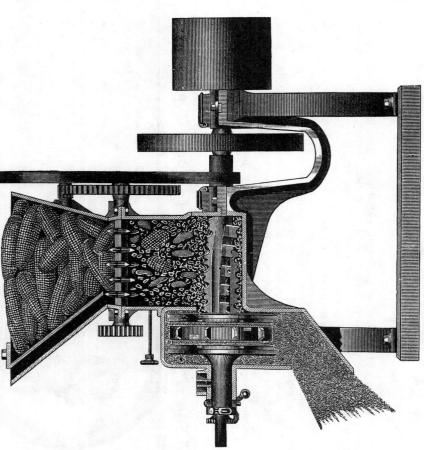

Nos. 1, 2, 3 or 4 Kelly Duplex, Arranged to Grind Ear Corn, Wheat, Oats, Barley, and all kinds of small grain.

It is provided with a double set of grinders or burrs, for which reason it is called the Duplex. It is the only mill manufactured which has double or duplex grinders. It has a grinding surface just double that of any other mill of equal size, and can, therefore, do twice as much work. It can do as much work as any other mill of double its size. It is very economical in the use of power. It will grind any kind of grain, cotton seed, corn and cob, corn, cob and shucks equally well, and will produce a grist of any desired quality, coarse, medium or fine. Its shifting device for regulating the grinding, is the most perfect known, and it is simple in construction, easily operated, strong, durable and efficient.

An extra set of four burrs are sent with each mill. We always ship the ear corn and fine grain grinders, as shown above, unless the shuck grinder illustrated on next page is specifically ordered. If shuck grinder is wanted it should be ordered at once as the attachment is not easily applied after the mill is sent out. The No. 1 size only is furnished with gearing instead of pulley when desired to attach direct to tumbling rod of horse power. Specification and prices on page 124.

A634—Osage, or Monarch Corn and Cob Mill.

Weight, 410 lbs. Price............$19.00.

The Osage and Monarch Corn and Cob Mills have in past years proven popular candidates for public favor. In size a strictly No. 2 mill, and corresponds with other No. 3 mills. Are fitted with celebrated cast steel grinders and force feed.

By reference to cuts of "grinding parts" of mills it will be seen that we have a double force feed and shear cut, which carries the grain rapidly through the mill; also greatly assists in grinding corn with shuck. We claim this mill will grind corn with shuck on better than any other mill made without extra attachments.

They are warranted to grind faster, wear longer and run easier than any mills of this class; grinds small grains superior to any other.

This mill is recommended for grinding corn in the ear, shelled corn and oats, barley or wheat, when mixed with shelled corn for feeding purposes only.

We have not recommended this mill for grinding oats alone, but here is the experience with this mill at a United States military post:

Messrs. Montgomery Ward & Co., Chicago:

DEAR SIRS—In answer to your letter of the 5th inst., I beg to inform you that the Monarch Grinder works to my satisfaction and does grind Colorado oats well enough for feed purposes even when unmixed. I hope it will continue doing so even after the grinder has worn a little. Yours truly,
EUGENE A. VON WINCKLER, Sedalia, Col., U. S. A.
September 12, 1893.

A636—Grinding Plates or Rings for Monarch Mill.

Cast Steel Grinding Rings. Per pair............$4.00
Diameter 24 inches.

A 624—RICHMOND COB CRUSHER.

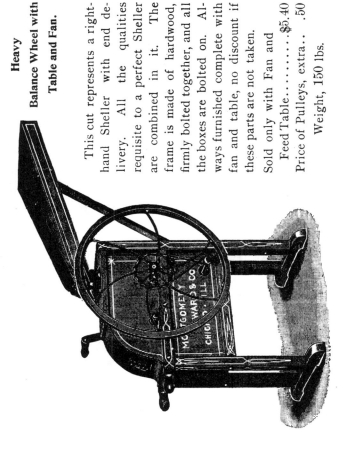

This crusher is only intended to crush corn in the cob and prepare the same to be fed into any of our feed mills before quoted. It is not recommended or guaranteed to make corn meal; would prepare coarse feed from corn in the cob for cattle. Reduces cob to about size of shelled corn.

DIMENSIONS.

The crusher requires a floor space of 16x32 inches

Height from floor to top of hopper, 26 inches. (The crusher may be set on supports at any desirable height from the floor.)

Pulley on crusher is 12 inches in diameter, by 5½-inch face, and is intended for a 5-inch belt. A larger pulley will be furnished, if so ordered, at a slight additional cost.

Speed: We recommend 350 to 400 revolutions per minute. The speed may, however, be varied, either slower or faster. Shipping weight about 240 lbs.

Capacity, twenty to twenty-five bushels of ear corn per hour.

This crusher in connection with any of our feed mills will constitute an outfit with which the owner can make money in almost any neighborhood.

Price (delivered on cars at Chicago).....................$16.50

A 632—"Eureka" Geared Feed Mill and Horse Power Combined.

The only Sweep Triple Geared Feed Mill in the market built with steel Pinions and Patent Adjustable Draw Bar Spring Attachment, taking the sudden jerking off the team when grinding.

Speed of tumbling rod to one turn of horses, 60 revolutions. One of our customers grinds from 12 to 26 bushels of mixed corn and oats per hour with his mill; another 14 bushels, and with his sheller attached, shells 35 bushels of corn at the same time. Power can be used for running feed cutter, circular saw, corn sheller, etc.

Mill and hopper can be removed if desired when used as power alone.

Weight, 700 lbs. Price........................$43.00
Burrs, per set..................................2.00

CORN SHELLER.

Heavy Balance Wheel with Table and Fan.

This cut represents a right-hand Sheller with end delivery. All the qualities requisite to a perfect Sheller are combined in it. The frame is made of hardwood, firmly bolted together, and all the boxes are bolted on. Always furnished complete with fan and table, no discount if these parts are not taken.

Sold only with Fan and Feed Table.........$5.40
Price of Pulleys, extra... .50

Weight, 150 lbs.

A758—HOCKING VALLEY TWO HOLE CORN SHELLER.

Weight, 225 lbs.

The Hocking Valley Two-Hole Sheller shown is intended for either hand or power. It is a Hand Feed Sheller, with a capacity of thirty-five to forty bushels per hour, and is admirably adapted for use in grist mills, and for farmers who are supplied with a power.

Sold always with Feed Table, Pulley and Fan...$10.80

HORSE POWERS.

All sweep horse powers furnished with two tumbling rods and three knuckles. Shipped as first-class freight.

A486—THE CAREY PITTS HORSE POWER.

NEW DEAL TWO AND FOUR-HORSE POWER.

This machine can be used with two, four or six horses.

Probably no other power is so well known to the farming community as the Pitts Power. It is one of those inventions that is born so nearly perfect as to be beyond rivalry or necessity of much improvement. It has always been noted as superior to all the rest for durability and ease of draft. Not only is the material of the best quality of air-seasoned white ash frame, and Salisbury charcoal (car wheel) iron gears, but the frame is stiffer and the gear heavier than in any other power of the same rated horse power; and this material is so distributed as to give the greatest possible strength for the weight.

This power is especially adapted for farm work where small power is needed, such as wood sawing, cider making, feed cutting and grinding. Diameter of main gear wheel, four feet. Measurement across power from one end of sweep to end of opposite sweep, twenty-four feet. Weight, 1,500 lbs. Number of revolutions, 52 to one turn of horses. Four sweeps, two tumbling rods and three knuckles go with each power, or one long tumbling rod with 16, 18 or 20 inch pulley attached, for running feed cutters, etc., which ever is preferred. Circular saws and grinding mills require a speeding jack.
Price.. $47.00

Hundreds of these powers in use and no call for repairs whatever.

A502—Four horse power, four levers, two tumbling rods, three knuckles, weight 1,100 lbs. Price $35.00. Speed 52 and 10½, or 15 and 10½, or 10½ and 4½, whichever is ordered.

One shaft furnishes the fast and the other the slow speed, as above stated. Diameter of main gear wheel 24 ins. Distance from end of sweep opposite, 24 feet.

This power has been thoroughly tested in all kinds of work for the past five years, and has given the best satisfaction. Just what every farmer wants to run his feed cutter, sawing machine, feed grinder, etc. Spring hitch on end of

A494—ONE-HORSE POWER.

Twenty-four revolutions to one turn of the Horses.

Ellwood's one-horse power......$18.00
(Including two tumbling rods.)

For running light pumps, cutting feed, shelling corn, churning, and other light work for which one-horse power is adapted. Weight, 400 lbs.

For pumping from deep wells, grinding and sawing wood, use power No. A486.

HORSE POWER JACKS.

A512—Spur Jack. A514—Bevel Jack.

Weight, 100 lbs. Price, $9.00. | Weight 100 lbs. Price $9.15.
14 cogs in pinion, 48 in spur wheel. | 20 cogs in pinion, 56 in bevel wheel.
Increases about 3½ times. | Increases 2¾.

Size of shaft squared to 1⅜-inch.

The Spur Jack transmits motion by belt at right angles to the tumbling-rod, and the Bevel Jack in the same direction in which the tumbling rod extends from the power.
Either Jack is strong enough for two or four horse power.

EVEN-TREAD RAILWAY HORSE POWER.

We furnish five sizes of Tread Powers as here described. Can supply single or double geared powers, but recommend the double gear for the following reasons: The double gear power is as prompt and responsive as any single geared power made, consisting of a double chain of lag-irons or links, with cogs, so constructed that *nine* teeth are constantly in mesh with those of the large gear-wheels on the main arbor, by which means a *perfect gearing* is secured. There is no possibility of lost motion. Any slack in the bridge or chain can readily be taken up by turning four nuts, thus the use of one-eyed or half links being dispensed with.

The speed regulator insures a motion as positive and uniform as that of a steam engine. The machines are both longer and wider in the tread than most others. The friction rollers are six inches in diameter. Average speed of band wheels, 103 revolutions per minute. Size, 4 feet, with 4½-inch face.

As to Level-Tread versus Even-Tread bridges the question simply resolves itself into this: whether it is easier to ascend an incline cut out in the shape of steps, or one that presents a perfectly smooth, even surface. We think almost any person would decide in favor of the smooth incline if he had to make the ascent himself. We have patterns for the level-tread, and can furnish it if deemed advisable to do so. We prefer, however, to build a power that we know will prove the most satisfactory in the long run.

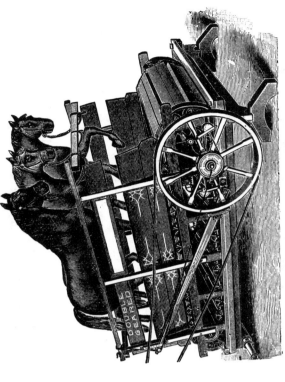

A526—BABY POWER.

For bull, calf, donkey, horse, or other animal, weighing not to exceed 1,000 lbs.
Price, including Speed Regulator and Pulley, $60.00.
Weight, 800 lbs. Length of Tread, 6 feet. Width of Tread, 15 inches.

A528—ONE-HORSE TREAD POWER.
With Speed Regulator.

Weight............1,200 lbs. Price............$75.00
Trucks for one-horse power. Weight 500 lbs.....$37.00

A534—TWO-HORSE TREAD POWER.
With Speed Regulator.

Weight............1,800 lbs. Price............$87.00
Trucks for two-horse power. Weight 600 lbs.....$40.00

A544—THREE-HORSE TREAD POWER.
WITH SPEED REGULATOR.

Weight............2,200 lbs. Price............$115.00
Trucks for 3-Horse, 630 lbs........................$47.00
Usual Diameter of band wheel, 4 ft, 4½-inch face, speed about 103 revolutions per minute.

A550—Four-Horse Tread Power, mounted on trucks, with Regulator and Raising Jack. Price...$205.00
Weight............4,700 lbs.
This tread power is about equal to an eight-horse sweep power and will run our large sized thrasher to its full capacity. Useful for Water Elevators and Pumps. Speed about 103 revolutions; Band Wheel 4 ft. 6 in. face.

THRASHING MACHINES.

A556—Thrashing Machine to be used in connection with three or four horse tread power. Is as complete a separator as there is made. Will thrash and clean all kinds of grain. Warranted to work as well as any machine made. Cylinder iron, with iron concave and sides. Size, 18x32. Weight, mounted, 2,650 lbs. Price, $130.00. Tailings Elevator, extra, $20.00. Straw Stacker, 10 feet. $30.00. Trucks, $40.00.

A558—Thrasher, and cleaner, same as above, but lighter. Cylinder, 18x28 inches. Weight, mounted, 2,000 lbs. Price, $100.00. Tailings Elevator, extra, $16.00. Straw Stacker, $1.50 per foot. Trucks, $40.00. Cylinder, with teeth boxes and pulley alone, $28.00.

A559—THRASHER AND SHAKER.

The Thrasher and Shaker, as shown in the above cut, is a very convenient and cheap machine, especially where Cleaner is not desired. They do good work, run light, and are built in three sizes.
Price, with 16x23-inch Iron Bar Cylinder............$65.00
" " 18x28 " " " " $75.00
" " 18x32 " " " " $80.00

A769—HOCKING VALLEY TWO HOLE CORN SHELLER.

The above cut illustrates the Hocking Valley Two Hole Hand Feed Sheller fitted for the direct attachment of the tumbling rod of a horse power.

A two-horse power is required, and the shelling wheel shaft of the sheller should make 125 revolutions per minute. We can vary the size of these sprocket wheels when desired, to accommodate the sheller to any power. All that is necessary to know is the number of revolutions the tumbling-rod makes to one round of the horses. It is furnished, as desired, either with or without Cob Carrier, Double Sacking Elevator, or Wagon Box Elevator. Weight, 325 lbs.

Price of Sheller as here shown with Feed Table, Double Sacking Elevator, Cob Carrier, and Tumbling Rod Attachment..............$35.00
Wagon Box Elevator in place of Sacking Elevator, adds to price. 5.00

A762—Hocking Valley Self Feed Corn Sheller.

The above cut shows the Self Feed Sheller with Double Sacking Elevator and Pulley. It is intended for use in mills, warehouses, grain elevators and other places where a belt sheller is wanted. It is furnished either with or without Double Sacking Elevator, Wagon Box Elevator or Cob Stacker, as desired.

This sheller has a capacity of sixty to seventy-five bushels per hour. The ears of corn are shoveled into the Feed Hopper and by means of sprocket chains the corn is carried up to the throat of the sheller.

It is made either as a belt or geared sheller. Speed of shelling wheel shaft should be 125 revolutions per minute.

In ordering, the size and speed of the drive pulley on power intended to run it should be given, so that a pulley of proper size may be sent with the sheller. All Self Feed Shellers are furnished with Cob Carriers. Weight, 375 lbs.

Price, with Feed Hopper and Pulley..............................$43.00
Price of Wagon Box Elevator, extra.............................. 15.00
Price of Double Sacking Elevator, extra.......................... 11.00
Price of Cob Stacker, extra..................................... 7.50
Direct Tumbling Rod Attachment in place of Pulley, extra........ 7.00
Cut illustrates machine with Feed Hopper and Pulley, double Sacking Elevator and Cob Carrier. Price as shown....................$58.00

The Improved Cummings Cutter, Nos. 3 and 4.

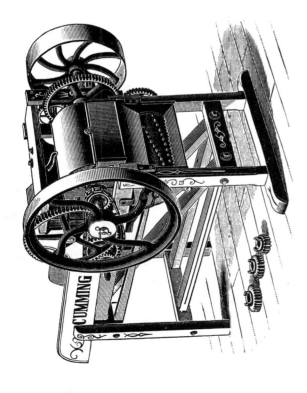

A 834—No. 3 is our popular Feed Cutter, the capacity being great enough to cut feed for forty to sixty head of stock. The No. 3 machine has been used to advantage as an ensilage cutter, and for small silos it will do very satisfactory work. Reversible carriers of any desired length are furnished when ordered. Can be used with or without fly wheel, and pulley can be placed on *either end of the shaft.*
Length of knives, 12 inches (always furnished with four knives); cuts four lengths—¼, ½, ¾ and 1 inch. Size of pulley, 14 inches diameter, 5½ inches face. Speed, 400 to 600 revolutions. Power required, two-horse tread. Capacity, 3,000 lbs. dry feed per hour. Weight, 500 lbs.

List Price	.... $70.00	Our Price $37.00
List Price, with Crusher	.... 95.00	Our Price 52.00
Price of 12-foot carrier (or under)		.... 16.00
Over 12 feet, extra per foot		.... 1.00
Knives extra for repairs. Weight, 3 lbs.		.... 1.25

A 836—The No. 4 is intended for power only. It is a very *heavy, powerful* machine, and is intended for large farmers, stock raisers, street car and livery stable men, and owners of silos.

It will cut feed of all kinds, and is the size generally ordered for ensilage. In construction it is similar in every respect to our other cutters, only heavier, larger and of greater capacity. Length of cut, ¼, ½, ¾ and 1 inch. Power required, three to six horse. Speed, 400 to 600 revolutions per minute. Capacity, about five tons per hour. Weight, 800 lbs. Length of knives 16 inches. Number of knives, 3 or 4. Size of pulley, 14 inches in diameter, 5½ inches face. We can furnish any size pulley desired.

List Price, with one Pulley .... Our Price	$120.00 $62.00
List Price, with Crusher .... Our Price	150.00 80.00
Carrier, 12 feet long or under; list price .... Our Price	30.00 16.00
Over 12 feet long, extra per foot; list price .... Our Price	1.60 1.00

A860—FODDER SHREDDER.

Patented June 13, 1882.

The above engraving represents a new and superior method of tearing up or shredding corn fodder, straw or hay, which is vastly better than the usual way of cutting.

It is not claimed that it excels in quantity, but in the quality of its work it has no equal.

As will be seen, it is a set of saws (instead of knives) placed upon an arbor in an angular position, making it one of the most effective machines for the purpose ever invented. When the fodder has passed the saws it is left in the best possible condition, as it is torn into irregular fragments, so fine that there is no waste when fed out. The hard ends or stiff stubs, liable to cause sore mouth when cut with the ordinary cutter, by this process are avoided. So soft or spongy is the mass that it will absorb a large quantity of water, and by mixing with shorts or meal, make a complete food for stock. The farmers have long felt the need of a cutter which would better prepare their coarse fodder. The Shredder supplies this want.

Stalks with ears on can be passed through this machine and excellent results obtained.

Price for No. A (2 to 3 horse power), ¾ tons per hour, weight 450 lbs.	$65.00
" No. B (4 to 6 "), 1¼ " 500 "	75.00
" No. D (10 to 12 "), 3 to 4 tons per hour, weight 900 lbs.	95.00
For No. 1 and No. 2, 12-foot carrier .... $22.00; for greater length add $1.20 per foot.	
For Jumbo, 12-foot carrier .... 24.00; for greater length add $1.20 per foot.	

A832—Improved Cummings Cutter No. 2
AS A HAND MACHINE.

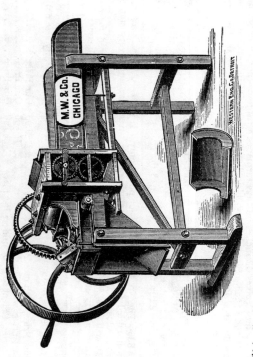

This Cutter has four 10-inch knives, is intended for hand power, and has no equal. It has the upward cut, the rocking feed roller and safety balance wheel. The power is applied directly to the knife-shaft which runs in two babbited boxes. Length of cut, ⅜ and ⅞. Weight, 300 lbs.
Price $16.00
Knives, extra each. Weight 2¾ lbs. 1.00
Prices named are correct. No reduction except our regular discount for cash with order.

This is a hand machine only. If wanted for hand and power, order

The New Dutton Knife Grinder.

Improved to date; the latest and best. But a moment's work to attach it to your mower wheel, and it is virtually self-adjusting. Anyone can use it. The water is placed in a core chamber, from where it works out automatically to the grinding surface.

Its use is a great saving in time, labor and wear of your mower.

Price, $9 00.

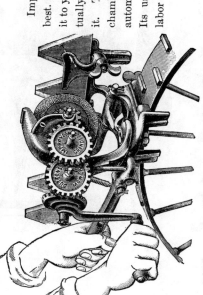

Our Special Nos. 3 and 4 Cummings Ensilage Cutter.

With Left Delivery Angle Carrier.

A 842—Price of 12-foot carrier; will work with any of our cutters...... $16.00
Longer, per foot, extra.. 1.00

A 844—THE CLIPPER CUTTER.

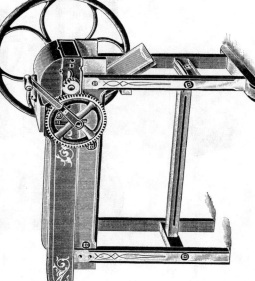

To satisfy a demand for a hand cutter, one adapted to general work upon hay, straw, etc., we brought out the above cutter a few years ago. We offer it to the public with perfect confidence in its superiority to any machine of the kind in the market, for rapid and easy cutting, large capacity and power, simplicity of construction, strength and durability, and moderate cost. This cutter is not intended to do the work of the largest horse-power cutter. When heavy corn stalks are to be cut we recommend the Cummings No. 1 or 2, as stronger and better adapted for heavy work.

It has one spiral knife geared to make four cuts to one turn of the crank, giving it a splendid cut at the least expense of power. Length of knife, 10 inches. The gears are few and strong, while it is almost self-feeding. The workmanship and material throughout are equal to that in our most expensive cutters.

List price............ $18.00. Our price............$9.50. Weight, 130 lbs.

A862—STOCK GROWERS' FEED CUTTER.

Weight, 700 lbs.

Price, with Elevator.....$115.00.   Price, without Elevator.....$95.00

This machine cuts corn fodder with ears on or off as fast or faster than one man can throw it off the wagon to the cylinder, the knives of which almost instantly cut the armful into the best feed.

Cuts sheaf oats, without unbinding, as fast or faster than one man can pitch the sheaves into the machine by hand or with pitchfork. Cuts hay, straw, peavines, etc., fed to machine with pitchfork. Cuts tobacco stalks and stems fed to the machine by the armful. Cuts ear corn, with shucks on or off, as fast or faster than two men can feed to the machine by hand or with grain scoops. Cuts turnips and other roots with the tops on or off, as fast as two men will feed them to machine with grain scoops.

In fact, machine has no equal, and is the most rapid, effective, complete and lasting cutting machine ever made, nothing to break unless, perhaps, occasionally a knife is broken, when a foreign substance, iron or stone gets into the machine; then a knife costing 20 cents, can be replaced in a few minutes. Machine does not have a cog wheel about it, consequently no wear or breaking of wheels or cogs.

The cutting knives are riveted on to a wrought iron cylinder and pass through a set of steel fingers, and so shaped as to make a draw cut, giving the easiest and greatest cutting capacity with the least power.

This Cutter cuts the stalks and ears of corn into suitable lengths; then by a set of chisel-shaped cleaners the pieces of stalks and cobs are crushed and torn to pieces lengthwise, and if the corn is dry every grain is shelled off the pieces of cob and the stalk reduced to a soft and spongy mass, making the feed well and easily masticated by the stock, and closely packs in silo when desired.

Speed of cylinder about 700 revolutions per minute. Power requires two or four horses.

A846—MONTGOMERY WARD & CO.'S CELEBRATED "CHICAGO" FEED CUTTER.

This is a very finely built machine that we have been manufacturing for several years, during which time all the small errors and imperfections incident to new machines have been discovered and removed, until the Cutter as it is now made represents as nearly a perfect machine of its kind as it is possible to make. We have spared no expense to turn out a strictly first-class article and honestly believe our Cutter stands without a rival. Selected wood stock is used throughout, firmly bolted together with wrought-iron bolts (no wood pins on our Cutters); the iron-work is prettily designed and strong enough to stand the heaviest work the machines can be subjected to, excepting accidents.

Made with knife 11½ inches long, quickly adjustable to cut four different lengths, ½ inch, 1 inch, 1½ inches and 2 inches. Will cut Hay, Straw and Cornfodder. Just the machine for parties keeping from four to five animals. Shipped knocked down. Weight, 165 lbs. Our price, $9.80.

A852—The Lever Cutter.

To those in want of a good, cheap Cutter for cutting for one horse or cow, we would recommend our Lever Cutter. It is made on the most approved pattern, with an adjustable gauge to regulate the length of cut. Has the best steel knife which can readily be ground, and a double tension nut with which to set the blade close up to its work. This is a standard and well known implement and is one of the best made of the kind. Weight, 60 lbs. Price.............................$2.80

A575—The Little Giant Green Bone Cutter.

FOR HAND POWER ONLY.

On stand as in cut. Price..........$9.00. Height, 45 inches. Capacity, 30 lbs. per hour. This Mill is adapted expressly for poultrymen and farmers, and is so low in price that anybody having use for one can easily afford to get it.

This Mill is so constructed that it will readily cut green bones, grizzle, and even sinews, direct from the butcher. The machine works on a shearing principle, the knife working directly against a steel cutter bar, which makes a thoroughly clean cut with every stroke of the lever. Weight about 66 lbs.

A576—Daisy Bone Cutter.

Price, with stand, 24 inches high....$15.00
This machine is especially made for cutting Green Bone with meat on right from the butcher, or offals of bone and meat from the table; also cuts vegetables, scrap cake, etc. A glance at the illustration shows at once the principle on which the cutter works. Knives operate like a common wood hand plane, except that they are notched and cut bone into small pieces. Weight about 140 lbs.

A583—Mann's Simplex Bone, Vegetable and Clover Cutter.

The Simplex is adapted for cutting clear meat, with or without bones. We would also call attention to the perfect automatic feed, which leaves both hands free to work with. It can be used as a clover cutter, thus giving the poultrymen a very necessary machine without extra cost. This will be appreciated by many who have not at hand all the labor-saving machinery of a large farm.

The Simplex machine is very strong, and will cut any bone you may put into it if you are willing to furnish the muscle. We put the machine against your strength, and will warrant it to stand the test. Weight, 38 lbs. Our price..........$9.00

Mann's Green Bone Cutters.

Green cut bone when fed to your poultry will double the number of eggs, will make the eggs hatch more vigorous chicks, and will increase the vigor of the whole flock. It will stop egg eating, will stop feather picking and will make little chickens grow faster and larger.

A thoroughly practical machine for cutting green bone, either by hand or by power. They are simple in construction, with nothing to get out of order, and easy to operate. They cut fast and fine. The knives (made of the finest steel can be taken out when dull, sharpened and replaced in a few minutes. You Can't Make Any Mistake if you buy a Mann Bone Cutter. They are strong and durable and will last for years.

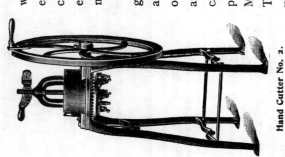

Hand Cutter No. 2.

A577—No. 2 Hand Cutter, like cut; weight, 142 lbs. Price......$16.50
A579—No. 10 Small Power Cutter, tight and loose pulleys, 2 balance wheels; weight, 147 lbs. Price.......... 23.50
A580—No. 14 Large Power Cutter; weight, 426 lbs. Price....... 68.00
A581—No. 15 Largest Power Cutter; weight, 720 lbs. Price..... 200.00

A582—Mann's New Departure.

To meet the want of those who desire a cheap yet a good Cutter, the manufacturers have produced the fine looking machine illustrated here. Don't think because it is cheap that it is no good. It is the result of years of study and experience in the bone mill manufacturing business. The knives and cutting parts are very carefully made and are the same as used on Mann's larger machines. This Cutter is intended for a flock of about 50 hens. Weight, 55 lbs.

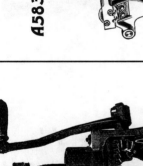

Price, only..........$7.00

Portable Farm and Dairy Scales.

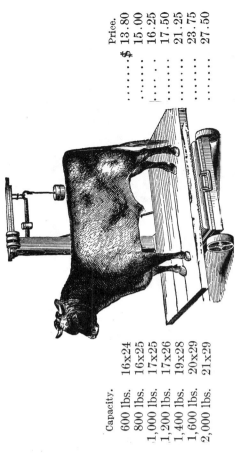

Capacity.		Price.
600 lbs.	16x24	$ 13.80
800 lbs.	16x25	15.00
1,000 lbs.	17x25	16.25
1,200 lbs.	17x26	17.50
1,400 lbs.	19x28	21.25
1,600 lbs.	20x29	23.75
2,000 lbs.	21x29	27.50

We furnish the very best make of portable farm scales, delivered free of freight to all railroad stations in the United States as far west as the west line of Kansas and Nebraska, North and South Dakota and Louisiana, at prices here quoted. To points further west we prepay freight to the amount of $1.00, and allow the purchaser to pay the balance on receipt of scales.

A674—COOK'S PORTABLE EVAPORATOR.

	Heavy Copper Pans.	Galvanized Iron Pans.
	$20.46	$14.63
	24.20	16.94

For the smaller pans, Nos. 1, 2, 3, 4 and 5, portable furnaces are provided. These are made of cast iron and sheet iron, heavily bound and riveted, strongly made, and the whole mounted upon rockers of angle iron—thus furnishing a complete portable furnace of iron and brick combined in one, with all the advantages of both, and yet so light that it can be easily handled by two men. This is the most convenient arrangement for small home operations, and for custom work it is well-nigh indispensable.

A796—The Hutchinson Cider and Wine Mill.

This is a well known implement. One of the most economical and useful Cider Mills manufactured.

Capacity, two to three barrels of cider per day. Weight 105 lbs.

Price..................................$7.50

A806—Allen's Broom Corn and Hay Press.

HAND LEVER POWER.

Weight, 1,300 pounds. Price..................................$74.00

This press has been tested satisfactorily for years by parties well qualified to judge of its merits, being practical growers and shippers of broom corn. They are in use in nearly every state in the Northwest. We also feel warranted in saying that its adoption is general, since it is used for a variety of purposes with modifications, and for rags, wool, hops, etc., it has no equal.

Bales of usual size, 30x24x45 inches, weight 275 to 350 lbs. of broom corn.

A444—SLIDING TABLE POLE SAW FRAME.

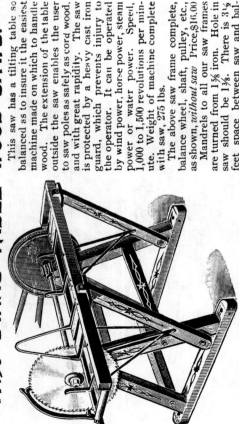

Our Sliding Table Saw Frame with Wheel below is built expressly to supply a long felt want in our trade for sawing poles, rails or any long wood as well as cord wood.

We have accomplished this by bringing the balance wheel below on brackets attached to the back of frame on which are bolted movable shaft boxes, which are set to tighten the belt from balance wheel pulley to the saw arbor; the extra expense, which is slight, cuts no figure when safety and convenience are considered. Weight 350 lbs. Price, *without saw, table alone*........$22.00

A446—SWING TABLE POLE SAW FRAME.

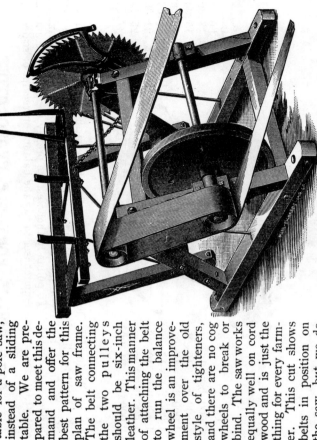

Some prefer a swing table for a pole saw, instead of a sliding table. We are prepared to meet this demand and offer the best pattern for this plan of saw frame. The belt connecting the two pulleys should be six-inch leather. This manner of attaching the belt to run the balance wheel is an improvement over the old style of tighteners, and there are no cog wheels to break or bind. The saw works equally well on cord wood and is just the thing for every farmer. This cut shows belts in position on the saw, but we do not furnish them unless specially ordered at an extra charge. All fully guaranteed. Weight 350 lbs. Price, *table alone, no saw*, $25.00.

A434—PRICES OF CIRCULAR SAWS.

Furnished to order on any of the following saw frames:

We quote price of *frame alone* and furnish the best Disston: 20-inch, $5.50; 24-inch, $6.60; 26-inch, $7.75; 28-inch, $8.80; 30-inch, $9.90. A 24-inch saw is as large as inexperienced parties should handle. Speed for all circular saws, 1,000 to 1,500 revolutions per minute. In filing saw, never cut sharp corners at base of teeth.

A436—SWING TABLE WOOD SAWS.

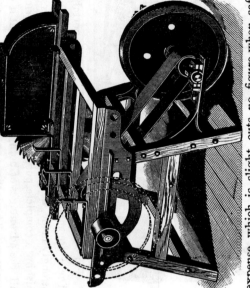

This saw has a tilting table so balanced as to insure it the easiest machine made on which to handle wood. The extension of the table outside the saw enables the user to saw poles as safely as cord wood, and with great rapidity. The saw is protected by a heavy cast iron guard, which prevents injury to the operator. It can be operated by wind power, horse power, steam power or water power. Speed, 1,000 to 1,500 revolutions per minute. Weight of machine complete with saw, 275 lbs.

The above saw frame complete, balance wheel, shaft, pulley, etc., as shown, *without saw* Price, $16.00 Mandrels to all our saw frames are turned from 1⅜ iron. Hole in saw should be 1⅜. There is 3½ feet space between saw and balance wheel, convenient to saw wood 5½ feet long into cuts. If longer, place stick on carriage diagonal so as to reach saw before touching balance wheel.

Price, Table, no Saw..........$16.00

Some think to save money by making their own frames. The Mandrel, two boxes, 85 lbs.; balance wheel and pulley, $10.00; if balance wheel is left off of frame or mandrel we deduct $3.00. Usual size of pulley or mandrel 2½ in diameter and 5½-inch face, larger furnished if desired. Price of balance wheel ordered alone, 5 cents per pound.

A442—SLIDING TABLE WOOD SAW.

Weight, 300 lbs. Machine ready to receive the saw without hood, as shown in cut, table alone, no saw, $17.50. Weight, 300 lbs.

For different sizes of saws to fit this frame see A434.

All of our saw machines are fitted with a balance wheel weighing 85 lbs. We furnish a heavier wheel weighing 125 lbs., if desired, at an extra charge of $1.50. Size of pulleys are: Diameter 5½ with 6½ face; 5½ with 5½ face; 5¼ with 5½ face; 12 inch with 6 face. Space between balance wheel and saw 3 feet 6 inches. Boxes babbitted. All work finished in the best manner.

No. C260—ALL STEEL FARM TRUCK

After persistent efforts and the expenditure of a large amount of money and time designing and building the special machinery necessary to build these gears, the manufacturers are enabled to place upon the market a steel gear which has been demonstrated by the large number in use during the past two years to be far superior to the old style wooden gears in strength, durability and lightness.

The axles are hollow and made with bolsters and stakes out of but two pieces of sheet steel. They are lighter than the wood gear and possess 25 per cent. more strength. Their main point of superiority, however, is durability—they will last a lifetime. The boxes, which are the only wearing parts, can be replaced when worn at a small cost. Will safely carry two tons. Thirty-inch front and 34-inch hind metal wheels with 4-inch bar-iron tire.

Guarantee.—Any axle or bolster proving defective in material or workmanship will be replaced free of charge.

Price, complete with tongue, as shown in cut $35.50
Price of Gear, without wheels 21.50

C636—OUR 4-KNEE BOB SLED.

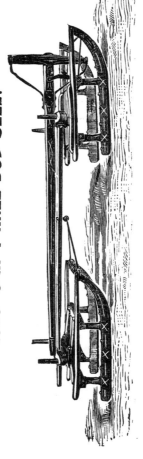

Weight about 325 lbs. Capacity, 3,500 lbs.
Price, only $9.99

Do not wait for snow before ordering, and lose more time in use than the cost of sled. DESCRIPTION.—4 knees, 2 in front and 2 in rear; extension reach; the tongues are oak; raves and reach, ash; bolsters, beams and knees, maple; runners, rock elm; size of runners, 1⅞ x 2⅜ inches; cast shoes, 40 inches in length; put together in the most workmanlike manner possible; thoroughly ironed and braced; painted two coats, striped and varnished; no whiffletrees or yoke furnished; the very best and cheapest bob sleds ever offered. Order early. The finishing painting, striping and ironing are first class in every respect, and equal to bobs we have always sold heretofore at a much higher price.

A1044—Kemp's Manure Spreader.

Covering every square inch with finely pulverized manure in one-tenth the time required with shovels and forks, and ten times as well.

It has a movable bottom composed of narrow slats or lags, connected with each other by a hinge chain, which moves on numerous small rollers. When the machinery is put into motion the bottom moves to the rear, its speed being regulated according to the amount of manure required to the acre, and brings the load in contact with a revolving toothed cylinder. This picks the material to pieces and distributes it evenly broadcast, or by the use of drill attachment, leaves it in two rows, as fast as the team moves along.

Above cut represents the Kemp Manure Spreader, with a belt beater instead of a round one. The advantages are that in light manure you can take a larger load, as the top of the beater is nearly ten inches higher than the sides of the machine, so that with the lighter grades of manure we have increased the capacity of the machine 50 to 75 per cent. With the very coarsest material you can load as high as the beater, and with the fine still higher. While you have the advantage of taking the large load of light manure you can handle the heavy grades of wet manure, another advantage is that by cutting straight through a deep load instead of in a circle much power is saved.

This machine is sold complete only, and mounted on its own wheels. A farm wagon or any part thereof cannot be used in this connection. No. 2, capacity 45 bushels; No. 3, capacity 55 bushels. Price and freight rates quoted to intended purchasers.

PRICE OF EXTRA ATTACHMENTS.

Drill attachment for leaving manure in rows (extra) $8.00
Wind Brake or Lime Hood (extra) 5.00
Lime board to convert drill to lime hood (extra) 1.50
Brake or lock for hilly sections (extra) 5.00
Slow feed for lime, ashes, cotton seed, etc. (extra) 1.50
Whiffletrees, Neck Yoke and Evener furnished free with all four-wheel machines.

A652—The Diamond Pumping Wind Mill Grinder.

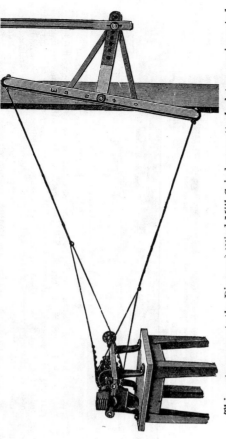

This cut represents the Diamond Wind Mill Grinder as attached to pumping wind mills. It is for the purpose of grinding feed for stock and making meal for family use. It will grind, in a good wind, one bushel of feed per hour. This Grinder can be located in any building anywhere from ten feet to one hundred and fifty feet from the mill. The hopper can be built over the Grinder, holding a quantity of grain, and when once put in operation will grind without further attention until the hopper is empty.

This Grinder can be operated with a ten-foot wind mill. By a new and practical device we are enabled to get a continuous forward motion of the shaft at both the upward and downward stroke of the connecting rod. The connecting rod being adjustable, the mill can be run in heavy or light winds.

Price, with three sets of burrs.................................. $15.00
Burrs, per set... 1.00

Connecting wire costs about six cents per rod in addition to price of mill and rock shaft. Full directions for setting up Grinder furnished purchasers.

Grinding Mill in Operation with Two-Horse Power

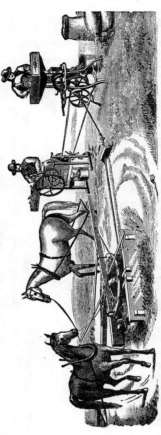

This cut represents our Big Giant Geared Mill Nos. 2 or 4 in operation with two-horse power.

No. 2 Mill A602......$23.75. Two-horse Power A496......$26.00. Horse Power $49.75
and Mill complete. Capacity, 10 to 25 bushels per hour.
No. 4 Mill A606......$32.75. Four-horse Power A502......$35.00. Horse Power $67.65
and Mill complete. Capacity, 30 to 35 bushels per hour.

A904—Diamond Pumping Wind Mill Attachment.

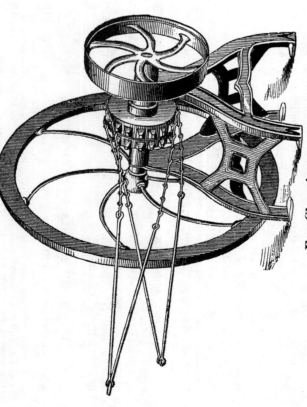

For Churning

This cut represents our device for transmitting the perpendicular motion of a pumping wind mill to a rotary. By using the same device that operates our wind mill grinder we are enabled to get a continuous forward motion of our shaft. Thus, by the use of a heavy balance wheel we are enabled to procure the rotary power to be driven by a perpendicular stroke of pumping wind mills. For driving light machinery, like churning, turning a grindstone, shelling corn, etc. Size of pulley, 12 inches in diameter, 2½-inch face. Size of balance wheel, 28 inches. Directions for setting up furnished.
Price, complete......$13.00

A 462—PRINCE SELF-FEED DRAG SAW.

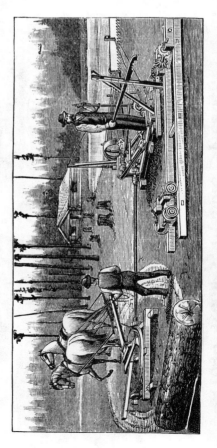

Weight, 1,100 lbs. Price, $56.00.

IMPROVED CIDER MILLS.

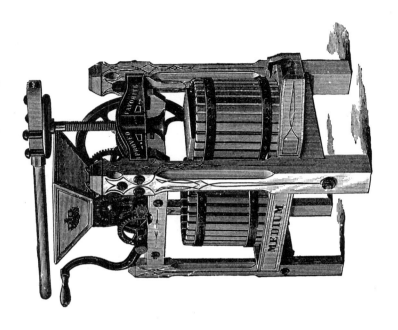

The Senior Mill has a screw 1¾ inch in diameter, and 26 inches long. It has the largest crates of any Senior Mill made. The frame is very heavy and strong. It will grind more apples in a given time than any Senior Mill manufactured.

The Junior Mill is a medium sized mill, screw 1¼ inch diameter, and 20 inches long. Is designed as a family mill, where one makes his own cider at convenient times. Where cider is made for sale, or by the gallon for other parties, order the Senior Mill.

A776—Senior.   Two Cranks and Two Crates.   Weight, 410 lbs.
Price..$20.50
A782—Medium.   Weight, 230 lbs,   Price..................17.50
A784—Junior.   One Crank and Two Crates.   Weight, 165 lbs.
Price..13.75

A1003—Link Belt Box Water Elevator.

A glance at the accompanying cut will give a good idea of just what this Link Belt Box Water Elevator is. There is so little friction in the working parts, and the floats fit so closely in the box, that very little more power is required than that which is theoretically necessary for the work of lifting a certain weight of water to a given height in a specified time. When steam power is used, therefore, the steam is employed economically and to the greatest advantage. Can be operated by horses or by water power if desired.

The capacities are given in gallons per minute, and are based on a belt speed of 300 feet per minute. Reduction of speed will reduce capacity proportionally. We do not advise much greater speed than 300.

We furnish all irons, brackets and bolts, with directions for making wood work, where parties choose to save freight by furnishing their own timber. Sprocket wheels used on horse power and on top of elevator, $2.00 to $3.00. Driving chain, 25 and 48 cents per foot.

	Number of Gallons per Minute.	Steam Horse-Power for 10-foot Lift.	Price of Elevator Complete with Wood Work, but without Driving Machinery or Pulley.					Driving Chains per foot.	Weight of 10-foot Elevator Complete
			10-foot.	15-foot.	20-foot.	25-foot.	30-foot.		
No. B.	600	2	$48.00	$55.00	$62.00	$69.00	$76.00	$0.25	536 lbs.
			Price Buckets and all Iron Work, no Pulley.						Irons.
			40.00	46.00	52.00	58.00	64.00	.25	226 lbs.
			Complete with Wood Work, no Pulley.						Complete.
No. E.	1100	3½	79.00	84.00	99.00	110.00	119.00	.48	1500 lbs.
			Iron Work and Buckets, no Pulleys.						Irons.
			67.00	75.00	84.90	92.00	100.00	.48	650 lbs.

THE MONTGOMERY WARD & CO.'S PUMPING WIND MILL.
WOODEN WHEEL.

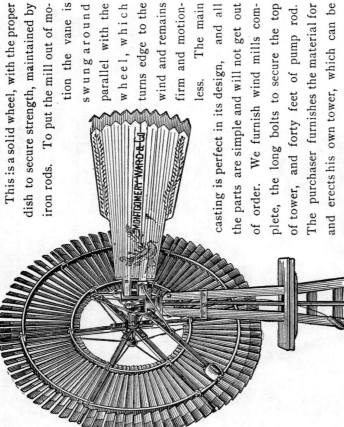

This is a solid wheel, with the proper dish to secure strength, maintained by iron rods. To put the mill out of motion the vane is swung around parallel with the wheel, which turns edge to the wind and remains firm and motionless. The main casting is perfect in its design, and all the parts are simple and will not get out of order. We furnish wind mills complete, the long bolts to secure the top of tower, and forty feet of pump rod. The purchaser furnishes the material for and erects his own tower, which can be done by any carpenter in a short time.

We furnish printed instructions for erecting, by which anyone can put them up. These instructions are very full in every particular, giving full description of work from beginning to end. The mill is so simple no one need fear any trouble in putting it in working order. We believe this pattern to be the best for a simple pumping mill, or to grind feed by means of a Rock Shaft or Tower Converter, connected to pumping rod.

A898—Price, complete, except tower, 10-foot mill; weight, 400 lbs....$24.40
A899—Price, complete, except tower, 12-foot mill; weight, 600 lbs.... 30.00

Shipping weight of mill, 574 lbs.

The cost of lumber and erecting the tower on ground should not exceed $10.00. Will furnish complete tower for 10 and 12-foot mills. F.O.B. factory.

A900—10-foot mill at 40 cents per foot.
A901—12-foot mill at 50 cents per foot.

We recommend a wooden tower for any kind of wood or steel mill.

A902—14-foot tower Wend Mill.

We furnish a 14-foot tower mill, with all gearing perpendicular and line shafting, that will give 2½ horse-power in an 18 mile wind with counter shaft for pumping, all complete, except tower, suitable for grinding, sawing wood, cutting feed, and ordinary farm work.

Price..$82.50

A903—THE ROYAL CROWN STEEL MILL.
FOR PUMPING ONLY.

8-foot Royal Crown Steel Mill, weight, 300 lbs; price..............$23.00
9-foot Royal Crown Steel Mill, weight, 325 lbs; price.............. 26.00
With this Mill we furnish an angle steel tower, price, per foot..... .60
Anchor Posts, per set of four..................................... 3.00

The six sections to this mill when bolted together and strongly braced in front, as shown, make a very rigid, strong and safe wheel. It will be seen that our wheel shaft having its two bearings eighteen inches apart, with the drive pinion just inside the outer bearing, is extremely well arranged for effective work and durability. An effectual brake is applied as the mill is pulled out, which absolutely holds the wheel still when out of the wind.

The vane is a single sheet of steel riveted to a bow-shaped bar of angle steel, the two ends hinging on the pivot casting. The construction of the wheel which is all steel, is such as to give the wheel a light, airy appearance, yet make it the strongest and most substantial. The arms, double bolted to the spider casting, are made from angle steel, being the strongest possible form for the weight of the material employed. The outer girths are made of channel steel bent on a circle, they are so shaped as to form a bearing the full width of the sail, thus strengthening the same and holding its curved form securely.

These results may be expected with this mill, velocity of wind, 12 miles per hour:

		25 ft.	50 ft.	75 ft.	100 ft.	125 ft.	150 ft.
8-foot Mill	Water Elevation	25 ft.	50 ft.	75 ft.	100 ft.	125 ft.	150 ft.
	Diameter of Cylinder	3½ in.	3¾ in.	3 in.	2¾ in.	2½ in.	2½ in.
	Gallons per hour	432	374	316	268	220	177
9-foot Mill	Water Elevation	100 ft.	125 ft.	150 ft.	200 ft.		
	Diameter of Cylinder	3¼ in.	3 in.	2¾ in.	2½ in.		
	Gallons per hour	374	346	268	220		

A1016—LIVELY AGRICULTURAL ENGINE.

PRICE AND WEIGHT.

6 Horse Power, weight, 4,900 lbs.	Our price	$436.00
8 Horse " " 5,800 lbs.	"	491.00
10 Horse " " 6,240 lbs.	"	540.00
15 Horse " " 8,230 lbs.	"	617.00

A 718—Montgomery Ward & Co.'s Road Grader.

Intended to be used on the gears of an ordinary farm wagon.

Weight, independent of trucks, about 1,500 lbs. **Price, without trucks, $65.00.**

Farm Wagon Gears, Whiffletrees, Stay Chains and Yoke adapted to use with this Grader: 3¼-in. Skein, $30.50; 3½-in. Skein, $32.50.

We guarantee this Grader to be strong enough to stand the strain of as many horses as it may be desirable to attach to the same, and all the parts easily operated and readily understood.

The draft on the Scraper Blade is direct through the Tongue, Hounds and Short Reach.

The Scraper has a strong bail and an upright iron sustaining post or shaft, and is pulled direct against the soil and not pushed from the rear.

This Grader Received Highest Award at World's Columbian Exposition, 1893.

APPENDIX

FARM MACHINERY CAPITALIZATION, 1908 vs. 1938

The two columns below represent the field tools needed on an average 160-acre farm in the years 1908 and 1938.

These implements were expected to last for twelve years if they were adequately housed.

Their acquisition cost in 1908 was about $925 and they could be purchased on a time payment plan. The value of the New York farm's 7-year old equipment in 1938 was $3,000.

A farmer's constant battle was to pay his property taxes and the 6 percent interest due on his revolving bank loan.

(Iowa) 1908	(New York) 1938
1 Grain Binder	1 Grain Binder
	1 Thresher
	1 Fanning Mill
1 Mower	1 Mower
1 Gang Plow	1 Sulky Plow
1 Walking Plow	2 Walking Plows
	1 Reversible Plow
	1 Tractor Plow
1 Riding Cultivator	2 Cultivators
1 Walking Cultivator	1 Cultivator
1 Disk Harrow	1 Disk Harrow
1 Smoothing Harrow	1 Tractor Harrow
	2 Spring- Harrows
	1 Steel Roller
	1 Weeder, 1-horse
2 Farm Wagons	2 Farm Wagons
	3 Bobsleds
1 Light Road Wagon	1 1931 Chev Truck
	1 10-20 Tractor
1 Corn Planter	1 Grain Drill
1 Grass Seeder	1 Grass Seeder
	1 Potato Planter
	1 Potato Hiller
	1 Potato Digger
1 Manure Spreader	1 Manure Spreader
	1 Lime Sower
	1 Marker
1 Hay Loader	1 Hay Loader
1 Hay Tedder	1 Hay Tedder
1 Hay Rake	1 Dump Rake
1 Buggy	1 1931 Pontiac

AVERAGE WAGES OF FARM LABOR, 1866 – 1902

Year	Per month		Per day		Per day during harvest		Difference per day with and without board
	With board	Without board	With board	Without board	With board	Without board	
1866	$17 45	$26 87	$1 08	$1 49	$1 74	$2 20	$0 46
1869	16 55	25 92	1 02	1 41	1 74	2 20	46
1875	12 72	19 87	78	1 08	1 35	1 70	35
1879	10 43	16 42	59	81	1 00	1 30	30
1882	12 41	18 94	67	93	1 15	1 48	33
1885	12 34	17 97	67	91	1 10	1 40	30
1888	12 36	18 24	67	92	1 02	1 31	29
1890	12 45	18 33	68	92	1 02	1 30	28
1892	12 54	18 60	67	92	1 02	1 30	28
1893	13 29	19 10	69	89	1 03	1 24	21
1894	12 16	17 74	63	81	93	1 13	20
1895	12 02	17 69	62	81	92	1 14	22
1898	13 43	19 38	72	96	1 05	1 30	25
1899	14 07	20 23	77	1 01	1 12	1 37	25
1902	16 40	22 14	89	1 13	1 34	1 53	19

TABLE OF WEIGHTS AND MEASURES

Measuring Grain.—A bushel of grain contains approximately $\frac{5}{4}$ cubic feet. To determine the capacity of a bin, find the number of cubic feet and multiply by $\frac{4}{5}$, or multiply by 8 and divide by 10.

Measuring Ear Corn.—It requires about two bushels of ear corn to make one bushel shelled. To find the capacity of a crib, find the number of cubic feet and multiply by $\frac{2}{5}$ or $\frac{4}{10}$.

Measuring Hay.—The quantity of hay in a mow is very hard to estimate accurately. The deeper the hay is, the harder it will be packed. Some kinds of hay are heavier than others, the longer it stands the more compact it becomes. Settled hay will usually weigh about five pounds per cubic foot. Or, 400 cubic feet will weigh one ton.

Measuring Land.—The easiest way to calculate land measurements is to figure 160 square rods as one acre. A strip one rod wide and 160 rods long, therefore, equals an acre, as does a strip four rods wide and 40 rods long, or eight rods wide and 20 rods long, etc.

A surveyor's chain is four rods long. It is divided into 100 links, so that all calculations are in decimals. Ten square chains equal one acre.

SQUARE MEASURE EQUIVALENTS

Sq. in.	Sq. ft.	Sq. yd.	Sq. rod	Acre	Sq. mile
144=	1				
1,296=	9 =	1			
39,204=	272¼=	30¼=	1		
6,272,640=	43,560 =	4,840 =	160=	1	
	27,878,400 =	3,097,600 =	102,400=	640=	1

HORSE LABOR vs. TRACTOR LABOR, 1938

	Average per Horse	Percentage of Total
Costs		
1,853 pounds of grain at $1.40 per hundredweight	$25.86	20.2
3.2 tons of hay at $7.05 per ton	22.57	17.7
Pasture and fences	4.80	3.8
Other feed and bedding	4.20	3.3
Total feed and bedding	$57.43	45.0
95 hours of labor at 30 cents	28.40	22.2
Depreciation	18.90	14.8
Use of buildings	8.75	6.9
Interest on value of $148 per horse	7.40	5.8
Shoeing	2.78	2.2
Veterinarian and medicine	0.92	0.7
All other	3.13	2.4
Total cost to keep a horse	$127.71	100.0
Credits		
8.4 tons of manure at $1.07 per ton	$ 9.02	
Colts, fair premiums, and the like	1.05	
Total credits	$ 10.07	
Net cost of horse work	$117.64	
Harness cost	4.96	
Cost of 699 hours at 17.5 cents	$122.60	

Annual Costs of Operating Two-Plow General-Purpose Tractors Mounted on Pneumatic Tires
(On Iowa Farms, 1936–1937)

Number of tractors	61	
Repairs	$ 19.38	
Fuel, oil, and grease	178.03	
Labor to service	7.83	
Depreciation	119.13	
Interest	42.35	
Taxes	22.58	
Total costs		$389.30
Hours drawbar work	667	
Hours belt work	116	
Total hours	783	
Total horsepower hours	9958	
Cost per hour	$ 0.50	
Cost per horsepower hour	.04	

Annual Costs of Tractor Operation
(78 Tractors on New York Farms, 1938)

	Average per Tractor	Percentage of Total
794 gallons of fuel at 11 cents per gallon	$ 90.30	38.5
18 gallons of oil at 64 cents per gallon	11.56	4.9
Farm labor	8.69	3.7
Insurance	1.36	0.6
Interest on average value of $493	24.65	10.5
Depreciation	66.56	28.4
Repairs	20.00	8.5
Use of buildings	5.92	2.5
Grease and greasing	1.95	0.8
All other	3.60	1.6
Total cost, 478 hours at 49 cents per hour	$234.59	100.0

NUMBERS AND VALUES OF FARM ANIMALS IN CONTINENTAL UNITED STATES

	Average Total number	Average Total value	Average Value per head
Horses—			
1867-1876	8,122,847	$516,776,357	$63.19
1877-1886	11,022,680	693,368,517	62.67
1887-1896	14,640,702	859,623,091	59.51
1897-1906	15,787,407	877,903,759	54.05
Mules—			
1867-1876	1,175,543	92,287,376	77.66
1877-1886	1,788,987	128,281,822	71.02
1887-1896	2,280,411	158,260,797	69.55
1897-1906	2,602,373	176,754,293	65.25
Milch cattle—			
1867-1876	9,998,355	283,515,175	28.42
1877-1886	12,646,159	336,001,308	26.47
1887-1896	15,861,965	360,505,202	22.79
1897-1906	16,948,692	487,693,745	28.74
Other cattle—			
1867-1876	14,957,992	265,992,932	17.69
1877-1886	24,227,144	475,656,436	19.12
1887-1896	35,331,043	562,422,695	15.96
1897-1906	38,463,070	713,738,958	18.98
Sheep—			
1867-1876	35,714,438	80,586,544	2.27
1877-1886	43,756,701	97,979,426	2.23
1887-1896	43,652,314	94,192,051	2.15
1897-1906	48,866,599	134,085,793	2.72
Swine—			
1867-1876	27,761,442	126,707,584	4.58
1877-1886	38,821,536	196,704,251	5.02
1887-1896	47,219,664	237,864,737	5.04
1897-1906	45,512,764	265,059,503	5.72

STATISTICS SHOWING THE PROGRESS OF AGRICULTURE IN THE UNITED STATES

Year	Total Population	Number of farms	Total acres of farm land	Average acres per farm
1850	23,191,876	1,449,000	293,560,614	203
1860	31,443,321	2,044,000	407,212,538	199
1870	38,558,371	2,660,000	407,735,041	153
1880	50,155,783	4,009,000	536,081,835	134
1890	62,622,250	4,565,000	623,218,619	137
1900	75,994,575	5,740,000	841,201,546	147
1910	91,641,195	6,366,000	881,431,000	139
1920	105,273,049	6,454,000	958,677,000	149
1930	122,288,177	6,295,000	990,112,000	157
1940	130,962,661	6,102,000	1,065,114,000	175
1950	149,827,932	5,388,000	1,161,420,000	216
1990	248,718,301	2,140,000	987,420,000	461

BIBLIOGRAPHY

Magazines & Newspapers:
American Agriculturist, NY, 1843 - 1884
The Cultivator, Albany, NY, Feb.- Dec., 1850
Farm, Field, and Fireside, Chicago, Jan. - June, 1884
Farm Implement News, Chicago, 1888 - 1892
American Thresherman & Farm Power, 1916 - 1918
Country Gentleman, Philadelphia, 1916 - 1918
Farm Mechanics, Chicago, 1915 - 1930
Successful Farming, Des Moines, IA, 1917 - 1924
The Furrow, John Deere, Moline, IL, 1924
The Chronicle, Early American Industries Assoc., 1933 - 1973
Farm Collector, Topeka, KS, 2002

Trade Catalogs:
B. K. Bliss & Sons, Springfield, MA, and New York, 1875
U. S. Wind Engine & Pump Co., Kansas City, MO, 1885
C. Aultman & Co., Steam Engines and Threshers, Canton, OH, 1889
Buffalo Pitts, Engines and Threshers, Buffalo, New York, 1891
Frank Brothers, Agricultural Implements, San Francisco, 1892
Montgomery Ward & Co., Farm Machinery, Chicago, 1896
McCormick Harvesting Machine Co., Chicago, 1897
Adriance, Platt & Co., Farm Machinery, Poughkeepsie, NY, 1907
Sears, Roebuck & Co., Hardware Catalog, Chicago, 1903 - 1927
John Deere & Co., Farm Machinery, Moline, IL, 1912
Minneapolis Threshing Machine Co., Minneapolis, MN, 1914
B. F. Avery & Sons, Tillage Implements, Louisville, KY, 1915
J. I. Case Threshing Machine Co., Racine, WI, 1918
McCormick-Deering, International Harvester, Chicago, 1919 - 1926

Reference Books:
A Pictorial Encyclopedia of Trades and Industry, Denis Diderot, 1751
The New England Farmer, Samuel Deane, Worcester, Massachusetts, 1790
Gleanings from Books on Husbandry, James Humphreys, Philadelphia, 1803
The Compendium of Agriculture, William Drown, Providence, 1824
Panorama of Professions & Trades, Edward Hazen, Philadelphia, 1837
The Complete Farmer & Rural Economist, T. G. Fessenden, Boston, 1842
Farm Implements, Principles of Construction, John Thomas, New York, 1854
One Hundred Years Progress of the United States, C. L. Flint, Hartford, 1870
Great Industries of the United States, Horace Greeley, Hartford, 1872
Asher & Adams Pictorial Album of American Industry, London, 1876
Appleton's Encyclopedia of Applied Mechanics, New York, 1896
Farm Machinery & Farm Engines, Davidson & Chase, New York, 1908
Equipment for the Farmstead, H. C. Ramsower, New York, 1917
Operation, Care & Repair of Farm Machinery, Deere & Co., Moline, IL, 1930
Farm Management & Marketing, V. Hart and M. Bond, New York, 1942
100 Years in Agriculture and Industry, International Harvester, Chicago, 1947
American Science and Invention, Mitchell Wilson, New York, 1954
Country Storekeeping in America, 1620 – 1920, Laurence Johnson, Rutland, VT, 1961
A Museum of Early American Tools, Eric Sloane, New York, 1964
An Age of Barns, Eric Sloane, Ballantine Books, New York, 1967
Farm Tools through the Ages, Michael Partridge, Boston, 1973
Wind-Catchers, American Windmills, Volta Torrey, Brattleboro, 1976
Smithsonian Book of Invention, Russell Bourne, editor, New York, 1978
The Branding of America, Ronald Hambleton, Yankee Books, 1987
Victorian Houseware, Hardware & Kitchenware, Ronald S. Barlow, Dover Pub., 1992
Classic John Deere Tractors, Randy Leffingwell, MBI Publishing, 1994
American Agriculture, A Brief History, R. D. Hurt, Iowa State University, 1994
Encyclopedia of American Farm Implements, C. H. Wendel, Krause Pub., 1997
The American Tractor, P. W. Ertel, MBI Publishing, 2001
The Antique Tool Collector's Guide, Ronald S. Barlow, L-W Books, 2002

INDEX

Adriance, Platt & Co., 62
Aermotor, 87 - 88
Advance Rumely, 74, 131
Appleby, John F., 57
Aspinwall Mfg. Co., 84
Atkinson, C., 45
Aultman, Cornelius, 71, 114 - 115
Aultman & Taylor, 71
Avery, & Sons, 37, 47, 55, 74
axe, 7, 22, 26 - 27

Ball, E., 65
barrel cart, 177
bean harvester, 165
Bell, Patrick, 57
bee hive, 13
bob sled, 93, 197
boiler, farm, 177, 183
bone cutter, 194
broadcast seeder, 174 - 175
Brown, George W., 44. 46
Brown, H. I. & C. P., 43
buckboard, 144
Buckeye mower, 65, 114
Buffalo Pitts Co., 70, 116 - 117
buggy, 92, 150
butter churn, 14, 78 - 79

cart, 92 - 93
carriage, 93 - 94
Casaday, W. L., 36
Case, Jerome, 72, 118
Case, J. I. Co., 98 - 101, 112, 121
cast iron, 26, 33 - 34
Caterpillar Co., 74, 119
Champion mower/reaper, 56
check-row planter, 44, 46
cheese press, 16, 78
cider mill, 195, 199
Collins, Samuel, 26, 35
combine, 74, 97, 129
Conestoga, 93
Cooper, C. & G. Co., 113
corn, 7, 8
    binder, 76, 125
    cutter, 12, 47, 172
    grinder, 83
    harvester, 76, 106, 172
    husker, 75

knife, 76
lister, 45, 109
mill, 12, 186
picker, 76
planter, 44, 47, 108, 125, 128, 169, 171
sheller, 13, 69, 76, 184 187, 190
shredder, 75, 105
cotton
    gin, 77
    planter, 47, 108, 128, 171
cows, 7, 10, 78 - 79, 82, 205
cradle, grain, 19, 28 - 29, 74
cream
    can, 80
    separator, 81
Crozier, Wilmot, 122
cultivator, 45, 53 - 55, 106 126, 136, 141, 155 - 157, 165

dairy equipment, 19, 78 - 81
Davenport, F. S., 35
De Laval, Dr., 81
Deere, John, 34 - 35, 45, 53, 74, 121, 123 - 129, 131
Deering, William, 57, 60, 74, 76
dibble, 27, 43
draw knife, 27
drill, grain, 43, 128, 137
Dutton, Rufus, 65

Emery & Co., 16, 20
engine
    gasoline, 130 - 131
    steam, 72, 74, 96 - 99, 113 - 119
Ertel, Geo. & Co., 68
ensilage cutter, 75
evaporator, sugar, 195
evener, two-horse, 166

Fairbanks - Morse, Co., 120
fanning mill, 14, 70
Farkas, Eugene, 122
Farmall tractor, 120, 122
Farquhar, A. B., Co., 48
feed cutter/grinder, 83, 185 - 186 191 - 193
fence, 18, 22

fertilizer, 8, 48 - 49, 51
flail, 27, 70
fodder shredder, 191
Ford, Henry, 122
Fordson, tractor, 122
Foster, N., 43
Fowler, William, 113
Frick & Co., 96
Froelich, John, 121

Gammon, E. H., 60
garden cultivator, 54, 155
Gilpin Moore patent, 36
Gladstone, Thomas, 57
glove, husking, 75
grain
    binder, 57. 59, 61, 63, 110, 125, 127, 183
    cradle, 19, 28 - 29, 57
    drill, 43, 128, 137
    mill, 83, 185, 198
guano, 48, 51
gypsum, 51

Halladay, Daniel, 87
hammer, 28
hammer mill, 83
hand tools, 26 - 31
harrow, 38 - 42, 111, 125 126, 167, 168
harvester, 15, 57, 61, 63, 76, 97
hay, 51, 64
    carrier, 180 - 181
    fork, 28, 29, 67, 180 - 181
    knife, 28
    loader, 67, 127, 182
    mower, 58, 63 - 65
    press, 68 - 69, 182
    rake, 64, 67, 138, 140, 178
    spreader, 13, 64
    stacker, 64
    tedder, 64, 66, 178
Haworth, G. D., 44
header, 63
hoe,
    clamshell, 8
    iron, 28 - 29
    horse, 43
    rotary, 127

hogs, pigs, swine, 8, 205
Holland, J. K., 51
Hollister, J. T., 61
Holt, Benjamin, 74, 119
Hoosier Drill Co., 96, 107
Hoover, Prout & Avery, 84 - 85
horses, 22, 74, 82, 121, 205
horse power
    sweep-style, 15, 182, 188, 198
    treadmill-style, 16, 20, 24, 189
Howard, Charles, 34
Huber Mfg. Co., 113, 120 - 121
husking bee, 75
    glove, 75
Hussy, Obid, 57 - 59, 65, 114

International Harvester Co., 60, 104, 108 - 111, 120, 122, 131

Jefferson, Thomas, 9, 34

Ketchum, William, 58, 65
Keystone Company, 60
Kirby, A., 65
knife
    corn, 76
    hay, 28
    pruning, 31
Krause brothers, 51

Lane, John, 34, 35
Liebig, J. V., 51
Locke, Sylvanus, 61
lister, 45, 47, 109, 128

manure spreader, 48 - 49, 50 - 51, 105, 127, 197
Marsh brothers, 61
Matteson, D. C., 74
mattocks, 7, 28

McCormick, Cyrus, 57 - 60, 65,
McCormick-Deering, 74, 76, 105, 106, 110
milk bottle, 78
    can, pail, strainer, 80
milking machine, 78 - 79
Miller, Lewis, 65, 114

Minneapolis tractor, 102 - 103
Moline Plow Co., 36
mower, 56 - 58, 63, 65, 106, 114, 125, 127, 139, 183
mules, 24 - 25, 121
Murphy, E. C., 88

Nellis & Co., 67
Newbold, Charles, 33 - 34
Nichols & Shepard, 71
Norse, Joel, 35

Ogle, Henry, 53
Oliver, James, 34 - 36
Osborne, D. M., 60, 76
Owens, Lane & Duyer, 113
oxen, 7, 11, 22, 24

Palmer & Williams
P. & O. Co., 60, 107 - 109
Perry, Thomas, 87
pig trough, 176
Pitts, Hiram & John, 70 - 71
Planet Jr., 155 - 156
plow, 32 - 37, 40, 42, 60, 107 - 108, 127, 133 - 135, 158 - 164
potato
    digger, 84, 127
    planter, 84 - 85, 128
    sprayer, 85
pruning shears, 31

rake
    grain binder's, 17
    hand, 27, 29
    hay, 138, 140, 178 - 179
    horse, 13, 64, 66
    side, 109, 127
    sweep, 59, 63
Ransome, Robert, 33
reaper, 12, 56 - 58, 63
ridge burster, 41
road grader, 201
Rogers, D. B. & Sons, 67
Rockwell, D. S., 44
Roberts, Cryus, 71
Rumely, M., 71, 73, 120, 122
Russell & Co. 117

Sandwich Mfg. Co., 83
saw, 28, 31, 196
scales, 195
scythes & sickles, 28 - 29
Sears, Roebuck, 50, 52, 130
seed sower (also see drill), 12, 17, 42, 173 - 174
shovel, 28 - 29
sickle grinder, 184, 192
side delivery rake, 66 - 67, 109, 127
silo, 75
singletree, 166
sledge, 93
Smith, F. F., 35
Spaulding, George, 61
sprayer, 85, 173
sprinkler, 176
steam engine, 48, 72, 74, 96 - 99, 113 - 119, 201
steamer, farm feed, 177, 183
Studebaker Wagon Co., 95
stump puller, 13 - 14

threshing machine, 20, 70 - 74, 112
tractor, gasoline, 98 - 103, 120 - 125
traction engine, (see steam engine)
trees, felling, 7, 22
truck, 104, 197
Tull, Jethro, 42 - 43

water tank, 176
Waterloo Engine Co., 121 - 123
wagon, 92 - 95, 143 - 149, 151 - 153
Walker, Edward, 67
Walsh, Leonard, 87
Washington, George, 9, 25
Weber Wagon Works, 60, 80, 104
Wemple, Jacob, 71
Westinghouse, George, 48, 71
Wheeler & Melick Co., 48
Wheeler, Cyrenus, 65
whiffletree, 166
Whitney, Eli, 77
Williams Brothers, 69
Wilson, Mitchell, 97
Windmill, 48, 66, 88 - 91, 198, 200
Wood, Jethro
Wood, Walter A., 63 - 64, 183